# Little Crystals

## A Low Sodium Journey
## Through the Salted Land

### BRETT L. MARKS

# Table of Contents

# Introduction

This book is the personal story of my attempt to solve a large and serious problem: *how do I significantly and permanently remove most of the salt out of my diet, and do so in a way that is workable and sustainable and does not lead to misery and deprivation?* Everyone could benefit from some of the material in this book, but there are specific populations that have been identified by medical professionals as needing to take on this challenge and find a personal solution for it. For these populations, their health is already at risk or they are nearing such a point. The number of people in this category is in the many millions, and it is growing. The at-risk populations eventually include nearly all of us as we age beyond fifty, but also people at any age with diagnosed high blood pressure, heart disease, diabetes, kidney disease and numerous other medical issues. Studies have also noted that African Americans in particular are at higher risk for many of these diseases and should be especially focused on salt intake, striving for 1,500 milligrams (mgs.) of sodium per day. This is also my target for daily sodium intake.

The task may seem daunting, especially at the start. What to do first?  What to do next?  This book will help with the step-by-step changes needed in what to buy, where to buy it, what to cook, and so forth.  Not everyone lives in large cities with many grocery store choices to increase the likelihood of finding low sodium foods, or near specialty stores that might carry a greater proportion of harder-to-find low sodium items. I have included a number of online references in this book to help in the search for low sodium foods regardless of where the reader lives.  In addition to these specific references, many products can also be found at *amazon.com* and at other online market sites featuring low sodium products such as *hearthealthyonline.com.*  Other sites with both product information and the ability to purchase products can easily be found by using search engines such as *google.com* or *yahoo.com* and typing in the name or type of product sought, such as low sodium baking powder.

Not everyone who reads this book will be an accomplished chef, or has even worked with many recipes in the past.  Consider me to be one of these people as well.  Some of the recipes in this book are quite easy, while others are

more challenging.  Do not be discouraged by the techni-
cal process of making a particular dish, and keep reading
even if it seems that some of the recipes will be a difficult
challenge for your individual cooking skills.  In addition, most
people lead very busy lives and some may wonder if they
really have the time to make these kinds of changes in their
shopping and cooking habits.  I face many of the same con-
straints on my time.  So I provide as many time-saving suggestions from personal experience as I can throughout the
book in recognition of the practical realities that many read-
ers face.

There isn't one narrow path to achieving a success-
ful low sodium focus in your diet.  Along with the recipes,
there are countless ideas and tips in this book that can help
the reader make better choices from a sodium standpoint
or can offer small but beneficial changes to an existing dish
which you are already comfortable preparing.  Take some
kernels from this book and apply them to your life - your buy-
ing habits and your own recipes.  You can do this, but your
chance of success increases by taking all the help you can
get from those who have toiled ahead of you, and consider
this book like an *a la carte* menu from which you can choose
the ideas, recipes and approaches that work best for you.

# I.
# August, 2006
# Understanding the Problem

*I think I may have a brain tumor....*

I was sitting propped up on the couch in my living room at 3 am, very much unable to sleep. I took in the surroundings – the shadows about the room, the low hum of the kitchen appliances, the drone of the air conditioning system from somewhere above me. The moon offered just enough light through the windows to make out the furnishings and the contours of the room. I wanted to return to the bedroom and lie back down on my bed and sink into the waiting cool pillows, but I could not. Not that night. I had already tried several times, and it was just futile. I was having one of those nights. The pounding in my head was relentless when I laid flat with my head on a pillow. It only subsided when I sat straight up. And it subsided rather quickly when I was prone. What was going on inside my head?

I was very worried.

The quiet darkness all about me was soothing. The coolness of the room was comforting. I went outside once or twice for a change in scenery and walked around the back patio in the sultry Texas summer night, but this change did not help. The thought of sleep slowly overtook me and I returned to the couch to wait it out. After what seemed an interminable wait, I eventually drifted into a less than satisfying sleep curled up on the arm of the couch, propped up on its upholstered pillow.

\* \* \*

The morning after the latest sleepless night, I finally made an appointment with Doctor Leonard, and about a week later, I confronted my brain tumor worries head on. The doctor knew my history but it had been quite awhile since I had seen him. He updated the file history with many questions about general health, diet and exercise. He took my blood pressure reading, and neither he nor I was happy about the elevated numbers. But he calmly took notes while I sat there dreading every second of the visit. He particularly focused on the pounding in my head when laying down to sleep, and whether I could remember any meals that I had on nights when this happened. Frankly, I could not remember. There wasn't any obvious sign of alarm on his face, and his questions had puzzled me enough to ask him for the step-by-step process of how we would go about determining whether the pain indicated a brain tumor. He chose not to answer the question at that moment, and instead kept methodically scratching notes onto his pad. This made me even more nervous.

Doctor Leonard's scolding started by telling me that I had waited too long to see him under these circumstances. He was very concerned about my high blood pressure, even accounting for the "white coat syndrome" that I seem to suffer. He could treat high blood pressure with medication, just like with many other patients. But the nighttime headaches as I described them were a different matter. Based on the totality of my history, especially food and lifestyle choices, he suggested that I was also suffering from sodium sensitivity, a far less common malady than high blood pressure, possibly affecting 5 to 10% of the general population. I had never heard of sodium sensitivity, but my immediate reaction to his words was more focused on the fact that there was no mention of any possible brain tumor. I was exhilarated and felt the lifting of a giant burden. Relief washed over me in waves, and at that moment I felt ready to deal with whatever he told me I needed to do.

Doctor Leonard advised me very plainly that I would need to change my diet and lifestyle, and become much more physically active. I would also have to reduce my salt intake. I had suspected that I had high blood pressure – it was high at the last visit to the dentist, who had encouraged me to see my doctor about it. But overall, this seemed like relatively great news, and I was determined to take a positive approach to end the symptoms once and for all.

I asked him how low I had to go in terms of salt reduction. The answer was 1,500 milligrams of sodium per day, although at that moment I didn't really know what it meant in terms of changes to be made. He also advised me that it won't happen overnight, but that I should give myself thirty days to get there. I had no idea how much sodium I consumed. He explained that from what I had told him about my food preferences, I was probably consuming between 3,000 to 5,000 milligrams of sodium per day, perhaps more when eating out. A lot more sometimes. It's very easy to run up big numbers without even noticing it. If you only had one teaspoon of sodium per day from all sources, including natural sodium in your food, that would be about 2,400 milligrams. He wanted me to ingest a good deal less than that -- a little over one half teaspoon per day, from all sources.

The doctor expressed a bit of sympathy, as he admitted to his own struggle with salty foods. It seemed we were both fans of pizza and Chinese food! He went on to list cheese, lunch meats, and salty snacks generally, and eating out as well. As the list grew, the first little pangs of potential deprivation started to tentatively creep into my thinking, and this began to drown out the voice in my head telling me that I am not about to die and that I should remain positive!

I needed to start reading every label on every food item that isn't totally fresh and in its pure, natural state, and count the sodium very carefully. Doctor Leonard cautioned that he has lots of patients who try to do this but who ultimately fail to get down to the 1,500 milligram level. He is honest with all of his patients, admitting that this was not going to be a snap. It will take an earnest effort to bring sodium intake under control. And these changes were going to be permanent.

Next, Doctor Leonard moved to another possible condition associated with both high blood pressure and excess sodium. My kidneys needed to be tested for underperformance, including possible signs of kidney failure. He gave me a pamphlet on living with kidney failure, which startled me. Clearly, I was starting to lose some of my initial exuberance.

The good news in all this was, if the diagnosis is correct and I made the necessary changes, then I wouldn't have any more night headaches and I would be able to sleep just like before this ever started to happen. All I had to do was take the medications he prescribed and then radically change my eating habits and lifestyle – that's all.

Everyone talks about fat and sugar, or carbohydrates or even cholesterol. I have seen all of the books and diets which describe the evils of excess in these food components. But salt is rapidly becoming recognized as a major health issue as well. We add it to everything and it's hiding in most of the foods we buy. It is true that we do need some sodium in our systems to remain alive. But our bodies need only a small portion of what we typically consume, around 750 milligrams per day according to Doctor Leonard. So my new daily sodium limit is a good deal higher than the absolute minimum, but that was still not very comforting.

Doctor Leonard handed me the prescriptions with his scribbles and numbers. He also gave me a final piece of sobering advice -- the first thirty days of removing salt from your diet are the hardest. It won't be difficult forever, but it may seem like forever for the first few weeks. You may crave salt for awhile, and not like the taste of things without salt. But your taste for salt will have diminished once you get through the first month. You may be surprised at how much you can taste other flavors once the salt is out of the way. This is really a positive result of removing all that salt from your taste buds.

Further tests confirmed a considerable loss of kidney function, and I was given kidney support medicine as an additional part of the routine. I also would likely experience mild swelling from the blood pressure medication, so he pre-scribed a diuretic to help with this. And my cholesterol was elevated too. Another pill. It was a gloomy assessment. I couldn't believe that I had arrived at this point and did not realize what was going on. The kidney problem really irritated me – I got mad enough to make sure that the first month of salt reduction was a success. But it was not easy.

# II.
# Facing The Numbers

Heading home from the doctor, my thoughts turned to the task at hand. What was I going to do first? I bet everything in the house was full of salt. What a big change, and a permanent one according to the doctor. Wait until my wife hears about this! Why did I wait? I knew that something wasn't right for quite some time. No tumor, but I'm still going to hear about this for a long time.

I have always enjoyed working with numbers. I am comfortable with budgets, financial data and reports, and many different types of numerical or statistical information. Somehow, though, the thought of turning 5,000 milligrams into 1,500 milligrams seemed like a numbers task that could get the best of me. The only way to begin, I figured, was to take an accounting of where things stood. What was in the kitchen now, and how much sodium did it contain? What did I have to do to get my numbers all the way down to 1,500?  And what about those restaurant meals?  They might be the most difficult part of this process.

\* \* \*

I opened the pantry door, armed with paper, pencil and calculator.  It quickly became evident that this was going to require a lot of change.  Later on I would discover that not all of these changes would be so difficult, and that there were substitute products for nearly everything we consume.  There were techniques to make this bearable, but at that moment, I wasn't feeling the optimism that I had promised myself.  Every can, bottle, bag and box was examined, all of the common products that we have used for years.  The numbers added up so quickly that it was startling.

A favorite soup – 870 milligrams per serving

Marinara sauce – 460 mgs. per serving

Canned beans – 380 mgs. per serving

Chips – 290 mgs. per 1 ounce serving

Soy sauce (I knew this would be high) –

1020 mgs. per tablespoon!

Instant oatmeal – 250 mgs. per serving

Boxed seasoned grain dishes –

280 mgs. or higher per serving

Canned vegetables – 340 mgs. per serving

Salsa, spicy mustard, ketchup – all with many more milligrams than I had  thought, with the salsa and ketchup at around 200 mgs. per serving.  Some mustards were fairly low, like honey mustard, while others were around 100 mgs. per teaspoon.  But I use mustard by the tablespoon, not the teaspoon!

Boxed cereals ranged from 0 mgs. for Kashi™ Autumn Wheat to several hundred mgs. per serving for others.  We actually had a couple of no or low sodium choices that we were already consuming, which was somewhat comforting.  But it wasn't nearly enough.

There were items that had no sodium in our pantry: a few cereals, old fashioned oats, dry pasta, plain grains like barley or couscous, rice, cooking oils, vinegars and tea. A few items had very small amounts of sodium, like raisins and canned fruit.  Most everything else was typically in the hundreds of milligrams per serving.  Based on what we currently had stocked in the pantry, getting to the 1,500 mgs. per day level would not be possible.  Could I live on barley and fresh produce?  Did I really like the taste of lentils by themselves? I didn't think so. Did I have to say goodbye to Snyder's™ pretzels and Progresso™ soups, sometimes consumed together?  I didn't know, but I didn't have a good feeling about it.

My examination of the refrigerator led to much of the same findings.  Bottles and jars were generally loaded with sodium!  Olives, bottled salad dressing, pickles, stir fry sauces, barbeque sauce and other condiments were all too high to be practical in my new life. Many of these items were purchased because of their low fat content, but with no thought about their salt content.  Milk has 125 mgs. per cup – how about that?  But the biggest shock was bread. I can't taste the salt in bread, but virtually all commercial breads have between 100 to 250 mgs. per slice.  By the time you add lunchmeat, cheese and mustard, a sandwich can easily approach 1,000 mgs!  This was starting to depress me. "Maybe the mayonnaise is better," I thought to myself.  I turned that bottle around, and lo and behold – it was as bad or worse than several of the mustard choices at about 90 mg per tablespoon. I had been using low fat mayo thinking it was better for me!

What could I give up and what would go on my "do not disturb" list?  I didn't know.  Lunchmeat was a big problem, even though our choices were very low in fat and high in protein. The turkey breast was around 390 mgs. of sodium for a two ounce serving, and the roast beef was 440 mgs. The deli cheese was also high, around 200 mgs. for a one ounce serving.  Everything was higher than I had thought it would be.  How could I have paid so little attention to all of this sodium in our kitchen?  As a result, now I might have kidney damage.  How did this happen?

I now understand very clearly how this happened. At the time I was 49 years old, in a high stress professional job, and not exercising nearly enough.  I paid lip service to healthier eating, while in reality too much of everything was passing between my lips – sodium, fat, sugar, cholesterol, calories.  Fortunately, the night headaches that I experienced were the wakeup call I needed, but many people aren't lucky enough to ever get a warning sign.  For many, high blood pressure and related problems are silent stalkers, doing their long term damage without much notice, unless you happen to be seeing your doctor regularly, or until the body suffers a major setback.  And sodium sensitivity?  Who knew about that?

The doctor said that salt *per se* is not the enemy.  We must have it in our systems to function.  But too much of anything can be harmful, and the enemy really is *excessive* sodium.

My wonderful wife Marjorie was helping me render this accounting. She moved on to our smaller pantry which we were using as a broom closet and for pet supplies. I noticed that she was methodically clearing off the shelves, transferring the items into different parts of the kitchen and utility room.

"What are you doing?" I asked, thinking it an odd time to begin a cleaning project. Marjorie barely looked up from her efforts, but stated very clearly, "I am making you a low sodium pantry space so that you can get this done."

I took a long look at the emptying shelves in the small pantry and realized two important things – one, I was indeed going to need a dedicated space for the changes that were soon to come, and two, I sure did have a dedicated wife. I began to feel a bit more confident that I was going to be able to make this work.

# III.
# Making Changes

The first thirty days went about as the doctor had described. The lack of salt was very noticeable, and I was annoyed at times by how bland things tasted; although I was determined to stick with it. I was taking an accounting of where things stood and making tentative steps toward doing things differently. I returned to the nearby grocery stores with a whole new perspective and purpose. Where were the low sodium alternatives? How low were they? What would taste good and what would taste like cardboard? These early exploratory sessions were not very efficient. I discovered that you can read labels for hours, if you don't keep track of time. At first it was difficult to identify all of the alternatives available. One store would offer some products, while another one would offer a different range of products. Store brands might be the answer, but you would have to inspect each item individually to make sure.

It would take a number of trips to each store to ascertain who had what, and where to buy certain, specific items. Over time, I noticed that the options for low sodium alternatives were growing steadily, so keeping up would require more effort, but it would be well worth it.

I started making lists – scrawled on paper at first, and later on spreadsheets that gave me the item name, sodium and other nutritional information, and the stores where each item was available. I was finding things, and was pretty happy about that. Not everything was a repeat purchase, though, because some foods just aren't close enough to the original when you remove most or all of the sodium. Also, there was no good bread to be found anywhere. No sandwich loaves, bagels, rolls or buns with little or no sodium. Something had to be done about this.

So I did something very out of character for me. I bought a Cuisinart™ bread machine. I had never used such a contraption before, and I had also never made a loaf of bread in my life. But I wanted to try to create something that I wasn't able to find in any of the stores – a simple loaf of bread without salt, with decent flavor and texture. I also bought a "no salt added" cookbook and tried the bread machine recipe. What I got from this first unpracticed attempt was a square loaf of heavy, dense bread that did not slice well and was a bit too grainy for my taste. It really wasn't like the bread in the stores or the homey, attractive loaves that some people seem able to make on their own. But it was a start – I was making changes!

After many experiments over the next year, I did come up with a very good, inexpensive no sodium (less than 5 mgs. per serving) bread recipe that starts in the bread machine, but has its final rise in a traditional bread loaf pan or on a baking sheet (for rolls) before being placed in a regular oven. This was my first real success at making something to solve the sodium problem, and it gave me the confidence to overcome many other challenges by substituting and creating satisfying alternatives. Better yet, this is the most complicated recipe in this book. Once mastered, the rest of the recipes will seem easier by comparison. I am purposely placing this recipe in the front of the book, because it was a foundational item in my aggressive pursuit of the low sodium lifestyle. It's not critical that this bread recipe be mastered first, or even mastered at all. There are some substitutes for regular bread that I discuss in the next chapter that will be acceptable to some people for at least certain purposes. But this one accomplishment symbolized the fight to achieve something good that cheated the salt gods over their hold on my taste buds. And it worked!

# Salt-free Bread

## Tools:
- Bread machine set on white bread, medium loaf, light crust (about 3 hours). Remove bucket from machine
- 1 medium mixing bowl
- 7 small chef's bowls – glass or ceramic
- Spoon measures – tablespoon, teaspoon, ¼ teaspoon
- Pastry brush, measuring cup, board for working dough
- Baking sheets and parchment paper – for rolls or hamburger buns
- One pound loaf bread pan – to finish bread in the oven

## Ingredients:
2 ¼ teaspoons bread machine yeast – place in very small bowl, and bring to room temperature

2 Tablespoons milk – place in small bowl and bring to room temperature

2 ¾ cups all purpose flour – place in medium mixing bowl

1 Tablespoon and 1 heaping teaspoon vital wheat gluten – mix well into flour

2 Tablespoons sugar – place in small bowl

1 Tablespoon apple cider vinegar – place in very small bowl

2 Tablespoons Canola or other vegetable oil – place in small bowl

¾ cup warm water (bottled, or chlorine free)

Olive oil – reserve for brushing on dough

## Directions:
Place water, milk, oil and vinegar in bottom of bread machine bucket.

Add flour and gluten mix into bread machine bucket.

Add sugar on top of flour; then add yeast on top of sugar, and keep yeast dry until starting the machine.

Insert bucket into bread machine and start the cycle. Ensure all flour is incorporated into dough after first few minutes of the dough cycle. Run the cycle until baking is about to begin, probably around one and one-half hours. When at the point of baking, remove dough from bucket and turn onto a floured board or surface. Remove the bread machine's dough paddle from the dough ball.

For bread loaf pan – lightly oil sides of pan and then add one pound of dough into the pan. This will leave about a half-pound of dough for rolls. Gently press so that all corners of the pan have roughly even amount of dough. Lightly brush the top of the dough with olive oil. Cover loosely with cellophane wrap and put in warm draft-free location to let rise until over the top of the pan (about 45 minutes).

Pre-heat oven to 375°. If you let the dough rise inside the lower oven of a double oven, do not cover the loaf pan while rising; otherwise cover with a linen cloth. Bake for 20 minutes or until top is light brown. Remove from pan and place on a cooling rack. When completely cool, slice. An overnight cooling is best.

For sandwich rolls or buns – the dough ball will weigh about 1½ pounds, or approximately 24 to 25 ounces. For large sandwich rolls, cut eight 3 ounce pieces of dough and shape into a ball, or if making a loaf of bread, cut three 3 ounce pieces of dough from the remaining amount after placing one pound of dough in the bread loaf pan. Place on baking sheet lined with parchment paper, pressing down on dough ball to flatten it out to about ½ inch thickness. Using a smaller amount of dough, like 2½ ounces, will generate about 10 smaller sandwich rolls or hamburger buns. Place half of the rolls on each baking sheet, leaving at least one inch between all dough surfaces. Lightly oil tops of each roll.  Place uncovered in a warm draft-free location like an oven to let rise until the rolls are about the size of the intended finished product (approximately 45 minutes).  Use a light cover of cellophane wrap for a rise on the counter or other area where there could be a draft.  Pre-heat oven to 375°. Bake for about 12 minutes or until tops are golden brown. Remove from baking sheets and place on a cooling rack. When completely cool, slice in half.

**Nutrition Information:**
Serving Size: 2 slices (from 1 pound loaf pan)
Sodium: 2.5 mgs.
Calories: 150
Total Fat: 3.5 g
Saturated Fat: 0.3 g
Cholesterol: 0 mgs.
Carbohydrates: 24.4 g
Fiber: 0.7 g
Sugar: 2.2 g
Protein: 3.8 g

Serving Size: 2 slices (from 1½ pound loaf pan), or one 3 oz. sandwich roll:
Sodium: 3.7 mgs.
Calories: 224.5
Total Fat: 5.3 g
Saturated Fat: 0.4 g
Cholesterol: 0 mgs.
Carbohydrates: 36.6 g
Fiber: 1.1 g
Sugar: 3.3 g
Protein: 5.7 g

## Hints and Tips:

I prefer to use a one pound loaf pan, reserving enough dough for three sandwich rolls or four hamburger buns – I find this to be the ideal combination for this amount of dough. The size of each slice is about the same as found in a commercially prepared loaf of bread. If you decide to use all of the dough for making a loaf of bread, the 1½ pound loaf pan is the preferable size. However, since this dough rises so much, the individual sandwich slices can get quite large on their tops.

This same recipe can be used to make dinner rolls too. Reduce the weight of the dough balls to make a variety of sizes. A good roll size is 1½ ounces of dough. I use an electronic scale that provides readings in ounces or grams. There are 28 grams in one ounce. Using the scale helps to ensure that I achieve the size I am looking for, and that there is pretty good uniformity with each roll.

Making a combination of bread, sandwich rolls and dinner rolls is easy to try as well. You can also add herbs and spices to the dough, or coat the tops of rolls with different types of seeds – sesame, poppy, caraway, for example. If you are adventurous, you may be able to create all sorts of concoctions to suit your tastes. This is a versatile, light, tasty salt free bread recipe. It freezes and toasts very well.

Because there are no preservatives other than the vinegar, it is advisable to refrigerate after the first couple of days, placing in the freezer after 3 to 4 days in the refrigerator. It will keep a long time if properly wrapped and stored in the freezer.

I purchased a bread slicing board with a slatted slicing guide to provide uniform slices, and I also purchased a long bladed bread knife that cuts cleanly through the loaf and stays within the slats of the guide. These products are available online in many styles and price points, and can also be found in restaurant supply and kitchenware stores. Finally, I asked the owner of my local fish and seafood store if I could purchase a handful of his plastic bags and twist ties so that I could use them to keep the bread. With his obliging my request, they cost only a few pennies per bag.

I did not stop there. After several botched attempts, I was finally able to reproduce the same loaf of bread or sandwich rolls using the regular stand mixer with the dough hook attachment and not using the bread machine at all. This is the same exact recipe, and I even place the ingredients in the mixer's bowl the same way, ending with the sugar and yeast on top. When using a dough hook, keep the mixer on one of its lower speed settings, allowing the hook to mix the dough for about 12 to 15 minutes. Remove the dough ball and place it into a good sized, lightly oiled bowl and cover. Let it stand for 45 minutes in a draft free place, just like the instructions indicated for the bread machine dough. It will more than double in size. After the first 45 minute rise, gently punch the dough ball down close to its original size. Then let it have a second 45 minute rise in a covered, draft free environment. At that point, you will have the same 1½ pound dough ball to work with for bread, buns or rolls.

It is very satisfying to make these bread products from scratch, and, somehow, not using the bread machine makes it feel like an even bigger accomplishment. Either way, this is a key solution to a sodium problem that vexes most people who are trying in earnest to reduce sodium intake - the hidden sodium in bread.

\* \* \*

One change was not very difficult to make. I had always relied on fresh fruit and vegetables as mainstays in my diet. Having discovered their low sodium benefits, I placed even greater emphasis on them, and hesitated to use a canned, frozen, jarred or bottled substitute unless I was convinced that the sodium content was little changed from the original fresh product. Fresh produce will almost always be a good choice. You may hear words to the contrary. When some people say that celery is salty, or even carrots or beets, there is a consideration to be made before deciding whether to be influenced by this in making purchase decisions. While a few vegetables may be saltier than some others, they are still very low in sodium compared to all of the processed and manufactured foods that we willingly buy, even when they say "lower sodium," for example. One cup of chopped celery contains 80 mgs. of sodium. That is a lot higher than a cup of cucumbers, at 3 mgs. But there are so many health benefits to celery – low calorie, good fiber and antioxidants, to name a few, that a cup mixed into a salad that serves several people isn't worth discussing. It is too beneficial to worry about this particular sodium content.

Reduced sodium *junk food* is a different story altogether. You should be very careful about these products, which may still contain loads of unwanted fat, sugar and even salt. Even "lower sodium" healthier products, like soups and pasta sauces, are still likely to have hundreds of milligrams per serving, although reduced from the original super-salty version of the product. I have found that "lower sodium" generally is not low enough for me. "Low sodium" foods are a much better choice for a diet this low in salt. "No salt added" may or may not be a good choice – you have to read the label to see how much salt exists in the natural state of the particular product. "No salt added" does not necessarily mean "no salt." "No salt" products tend to have zero to 5 mgs. of sodium. They are wonderful for the diet, but many of these items do pose challenges for good flavor. It is that challenge which is tackled throughout most of the pages of this book.

Another helpful change was achieved with the continued assistance of my very supportive wife, Marjorie. We have a main refrigerator in the kitchen, and an older, "spare" fridge in the nearby utility room. After some shuffling of items, we ended up with a "normal" fridge in the main kitchen and Brett's low sodium fridge next to the dryer and across from the new home for the pet food and cleaning supplies. All of my new concoctions and potions were separated from the so-called regular food. I have two grown daughters who need to be able to raid the refrigerator without fear of the unintended consumption of anything low sodium! This division of victuals has worked very well in our family.

The remainder of this book contains explanations, recipes, and other useful information, all based on maintaining a palatable 1,500 mgs. per day average intake of sodium, just like my doctor ordered.  Many people do not need to reduce their sodium intake to this low level, even though it could certainly be beneficial to do so.  The U.S. Government has revised downward its recommended level of sodium for various populations, and the current highest recommended level for people not at risk is 2,300 mgs. per day, or about one teaspoon from all sources.  That's quite low compared to average consumption.  Clearly, while the warnings and recommendations to reduce sodium are increasing, the public's consumption habits have not yet followed suit.  This book may be used as a resource for anyone who would like to understand the principles behind sodium reduction, and to reduce their intake to a lower level, whatever that level might be.  Going from 5,000 to 3,000 mgs. of sodium per day would produce nothing but positive results for the typical salt encrusted person.  But if you have been given the sobering medical advice to reduce sodium intake to the 2,000 to 1,500 mgs. level per day, consider this book as the "crib notes" to get you started on your journey through this salty land.  Getting started seems to be the largest obstacle for most people.  It is overwhelming at first, and there is much to filter through and much research to be done.  Changing habits is usually not easy, and these habits may have been ingrained from a very early age.  The whole world seems to be bathing in salt, and North America is among the worst with so many highly processed manufactured foods readily available. All of our favorite world-inspired cuisines are traditionally high in sodium as well, such as Asian, Tex-Mex, Italian and so forth. Eating is a fundamental part of life and sodium is a necessary ingredient, yet that which sustains us can also be taken in quantities that will reduce the quality of our health and ultimately shorten our lives.

# IV.
# Our Daily Bread

In the previous chapter, I shared a bread recipe that can be turned into sandwich slices, dinner rolls or hamburger buns.  Over the past few years, I have also found several other bread products that are either no or low sodium, and are worth considering for the ease of purchase versus baking from scratch.  While good, none of these products will serve as a complete replacement for the bread made at home.  But these products are ready to go and therefore, very convenient.

Low Sodium Ezekiel™ bread is a no-flour, whole grain bread that is dark, earthy and tasty (at least to me), especially when toasted.  Untoasted, this bread may become an acquired taste.  Despite its name, the nutrition label indicates that it contains no sodium at all.  The packaging is an interesting read as there is a biblical connection to this bread.  I keep a loaf on hand at all times.  I have reached the point where I can dispense with baking bread for awhile and use this product instead. That may not be the solution for everyone. It makes a pretty good toasted sandwich, and I have even used it for stuffing at Thanksgiving. Ezekiel™ bread is sold frozen, typically in a specialty bread section of the store. I have found it at Whole Foods™ and a few other local supermarkets in Texas.

The common, humble corn tortilla gets a co-starring role in this chapter. I have looked at the nutrition labels for many brands of these little disks, and quite a few of them are extremely low in sodium, averaging about 10 mgs. per tortilla. They have many good uses in a low sodium diet. They are great for breakfast tacos with eggs, onions, peppers and low sodium cheese and salsa. You will not miss the salt with this combination, especially if you season the eggs well with salt-free herbs and spices. They can also be used for other Mexican style meals with combinations of meat, chicken, vegetables, beans, rice, guacamole, sour cream, salsa and anything else that you bring to the low sodium fiesta. Later on in this book, there are recipes that make all of these ingredients available and flavorful on a low sodium diet. These tortillas should not be confused with their flour cousins, which are typically larger and unfortunately, contain on average about 300 mgs. of sodium each. That is true for both the white and whole wheat versions. This is a hidden salt that can't be readily tasted; yet can help run up the daily sodium intake very quickly. In the past, I always preferred the flour tortillas, but have learned to appreciate the corny version instead.

Another good bread choice is California lavash bread, available at Whole Foods.™ This version of Lavash is a large rectangular sheet of very thin, flat bread, like a rectangular burrito. It can be used to make two roll-up sandwiches by cutting it in half after it is filled and rolled, each one containing 90 mgs. of sodium from the bread portion alone. While this is much more than the other choices, it provides a good alternative to most store-bought breads and makes the kind of sandwich that could be a hit at a party.

* * *

What good is bread without something to put on it? The most basic combination is bread and butter. For me, no salt butter is too bland for spreading on no salt bread, even though it has many uses in cooking and baking. My preference is Smart Balance™ Low Sodium Buttery Spread, with only 30 mgs. of sodium per tablespoon. In my opinion, this is a good tasting spread. Smart Balance™ makes several versions, but only the blue label version is so low in sodium. Warm dinner rolls with this spread taste excellent, and upwards of 90% of the sodium is removed when compared to the traditional bread and butter combination. It will also melt nicely on toast. I cook with it as well – it is an all purpose low sodium substitute for salted butter.

Other tasty low sodium spreads include Simply Jif™ peanut butter at 32.5 mgs. per tablespoon, light cream cheese with ranges from 50 to 70 mgs. per tablespoon, and many zero sodium items such as jams, jellies, preserves, fruit spreads and pure nut butters made from only ground peanuts, almonds or other nuts.

The choices for low sodium sandwich fixings are still relatively few compared to their salty cousins, but are becoming tastier and more prevalent. For me, a good sandwich may contain meat, cheese, vegetables and condiments. What was once a 1,000 mg. serving can easily and tastefully be accomplished in the 200 to 350 mg. range.

At the deli counter, the choices are fairly few. Depending upon the region of the country, there may be a 40 to 80 mg, 2 ounce serving of deli turkey, chicken, ham or roast beef. Since I don't eat ham, and cannot find the chicken offering, the deli counter is a very limited visit for me. Boar's Head™ makes a good 80 mg per serving rare roast beef. Whole Foods™ carries a 30 mg per serving turkey breast in a pre-sliced package from Heidi's Hens™. It is rather expensive at around $14.00 to 16.00 per pound in 2011. I also roast chicken and turkey at home, and a thin slice of home cooked poultry is also a good component for a sandwich. Most stores carry a low sodium fresh chicken or turkey (whole bird or just the breast meat) in the meat section, and I aim for 50 to 75 mgs. per 4 ounce serving. Another good meat source is low sodium canned tuna, which is typically 30 to 40 mgs. per half can. I make a good tasting tuna salad using Hellman's™ Canola Mayonnaise, diced celery and red onion, plus at times I add a little bit of diced baby carrots for added color and crunch. I have also used green herbs like rosemary, basil, thyme and oregano to complete it. A tuna salad made this way contains about 200 mgs. of sodium per can, and I would use one-half can for a sandwich, or about 100 mgs. I use the whole can for a topping on a garden salad when it is the only source of protein in the salad.

The selection of low sodium deli cheeses at the deli counter is also fairly limited. My favorite is Lacey Swiss, which is made by Boar's Head™ and a number of other brands including store brands. It typically has 30 to 35 mgs. per 1 ounce serving, melts well, and is not as Swiss-tasting as some other choices. It is quite mild and even somewhat nutty in flavor. Alpine Lace™ makes even lower sodium Swiss cheese, all the way down to 10 mgs. per serving. These cheeses taste more like traditional Swiss to me. For an occasional treat, there is a low sodium Muenster at around 75 mgs. per serving. It melts better than any of the Swiss versions, and is milder and creamier.

The good news about lettuce, tomato, onion and most any other vegetable is that the sodium for the amount placed inside a sandwich is negligble. The one exception would be pickles, which are very high in sodium. But there are no-sodium bread and butter chips, which taste pretty close to the real thing. They tend to be rather sweet tasting, though, and do have a moderately high amount of sugar per serving. I pour out the liquid when first purchasing these pickles and replace it with plain water. This helps to cut down on the sweetness if that is your preference. Kroger™ sells these pickles under their store brand. B&G™ also makes them but I have never found them in my local stores. I also make my own not-so-sweet pickles, the recipe to follow in a later chapter.

Finally, many of the condiments that help add layers of flavor to a sandwich are available in low sodium choices.

Plochman's™ Honey Mustard is a zero sodium product, while other honey mustards are typically the lowest available mustard choice and are under 30 mgs. per 1 teaspoon serving. There is a wide variation in other mustards, but many types are available at 50 mgs. per 1 teaspoon serving. I will not buy any mustard without a nutrition label, or where the product is above 50 mgs. per one teaspoon serving. Read the label! Also, be careful – the labels can and do switch from teaspoons to tablespoons, so make sure you check carefully. Most mayonnaise contains between 90 to 115 mgs. per tablespoon. This is a fairly high sodium condiment, but used sparingly, it can enhance the flavor of many items. I use the Hellman's™ Canola Mayonnaise since it has half the fat of the regular product and tastes very good at 90 mgs. of sodium per tablespoon. So despite being higher in sodium, mayonnaise is on my "do not disturb" list because of the flavor it adds to so many items. I measure it carefully to make sure that I get the biggest bang for my sodium buck. Combine no salt ketchup with the canola mayonnaise, and a low sodium Russian dressing can be made for sandwiches or for use as a salad dressing, resulting in about 40 mgs. per tablespoon. Given its great taste, this is a very good sodium investment. Boar's Head™ also makes a very tasty Pub Style Horseradish Sauce that contains only 15 mgs.

of sodium per teaspoon, and adds a touch of heat to any sandwich.

Here are a variety of sandwiches made by combining these no and low sodium ingredients:

**Using no-sodium white (home baked) or toasted Ezekiel™ bread**

- Two ounces of Boar's Head™ low sodium Roast Beef, 1 ounce Lacey Swiss, lettuce, tomato, onion, no-salt pickles, if desired, plus 1 tablespoon honey mustard. This sandwich can be as low as 120 mgs. of sodium or up to 250 mgs. by using other low sodium mustards, horseradish sauce or by using mayonnaise. Substitute the low sodium Muenster cheese instead of the Lacey Swiss cheese and the total rises to not more than 300 mgs.
- Two ounces of Heidi's Hens™ Turkey Breast, 1 ounce Lacey Swiss, lettuce, tomato, onion, no-salt pickles, if desired, plus 1 tablespoon honey mustard. This sandwich can be as low as 80 mgs. or up to 210 mgs. by using other low sodium mustards, horseradish sauce or by using mayonnaise. Substitute the low sodium Muenster cheese instead of the Lacey Swiss cheese and the total rises to not more than 250 mgs. The no-sodium bread and butter pickles are a nice addition if you are so inclined. This turkey is a little bland, but it is so low in sodium that the flavor enhancing items can easily be accommodated. This is a good candidate for the zippy flavor of the horseradish sauce.

When making these same sandwiches with the lavash bread, the total sodium count would rise by 90 mgs. per sandwich for the change in bread alone. I also spread a small amount of whipped cream cheese on one end of the lavash to serve as the glue to keep the rolled-up sandwich together. This adds another 30 - 50 mgs. of sodium per sandwich. I compensate for the higher level of sodium in the bread and cream cheese by not using Muenster cheese, and by combining zero sodium honey mustard with another spread, thus bringing down the average sodium count for the spreads. A great tasting, festive looking lavash sandwich can be served to unsuspecting guests for about 300 to 350 mgs. of sodium. I also like to use arugula as the lettuce of choice in a lavash sandwich. Sweet pickles counter the somewhat bitter green very well. I prefer Romaine lettuce or other broad leaf lettuces in a regular sandwich made on the sandwich breads.

For the corn tortillas, I tend to stick to the egg, cheese, onion, pepper and salsa combo, which is not really a sandwich. They could be used to make small sandwiches by folding in half after stuffing with many of the same deli ingredients. I always warm them up first, either in a warm pan on the stove, or if I want to melt the cheese, in the toaster oven on the bake cycle. A very fast snack consists of nothing more than a couple of corn tortillas and 1 slice of low sodium cheese, cut in half and placed on each tortilla, melted in the toaster oven or microwave with a small amount of a low sodium salsa. I prefer D. L. Jardine's™ peach salsa, containing 95 mgs. of sodium per 2 tablespoons, but I only use about 1 teaspoon per tortilla. This combined snack comes in at around 120 mgs. of sodium, yet it is satisfying in both taste and nutrition. There are many lower sodium salsas, some even lower than the peach that I use.

There is a trend that the reader should be noticing by now. I constantly experiment with combinations and variations in order to find as many good choices as possible. I mix and match almost everything in my low sodium inventory. This has helped produce a much more interesting long term diet plan with less repetitiveness, and therefore, less boredom. The goal is to improve the chances of sticking to the plan for good, and variation and experimentation have helped me achieve this goal.

There is another aspect to my quest which may be obvious from reading this book. I question all claims made on every product and by everyone selling low sodium products. That's why I stress careful review of nutrition labels for the quantity that constitutes one serving and specific data on its sodium content. One anecdote that should provide helpful advice to the reader involves the claim that a deli meat product has "no salt added." On several occasions, I have seen this statement used in connection with house prepared deli meat at a large grocery store chain. In each case, it's been true that the deli employees did not add any salt when roasting the meat. In fact, it is claimed that the product is "all natural." But these house-made products often do not come with a nutrition label, and when my tongue tells me there is salt in the product, it is usually correct. I have asked to speak to the manager, and asked to see the meat in its pre-cooked state. At times, this has meant a trip to the walk-in cooler for the manager, and examination of a cardboard box which contains the frozen meat prior to its final preparation in the deli department. Time and again, there is no nutrition label on the box, but there is often a statement along the lines of the following: *this product contains a solution of up to 18% water, salt and sugar.* I have seen some surprised managers when they discover the truth about their supposedly "no salt added" product. So this inquiry is often necessary, and it is what you must be willing to do to remove potentially hundreds of milligrams out of your daily sodium consumption.

# V.
# The Spices of Life

It is generally observed that salt makes food taste better. Great chefs are seen on the television cooking shows plunking handfuls of salt into their pasta water, or generously rubbing down their roasts with the little crystals before placing them in the oven. They constantly reiterate that food should be well seasoned, and they mean salt along with other seasonings. It's even becoming more popular on desserts, the "sweet yet salty" combination being irresistible to many. Salted caramels are but one example.

It's true – most foods do taste better if they are salted. But after my initial 30 day period of salt deprivation, with time for reflection and a little creativity, it became clear to me that salt is but one of many flavor enhancers. It may be the best, or the easiest way to kick up the flavor, but the rest of my life will be an exploration of all other possible flavor enhancers to take the place of those little crystals, and to hopefully do it very well.

I have discovered many good alternatives for enhancing flavor.  By category, they can be roughly grouped into:

- spices and herbs
- citrus or sours
- heat and smoke

Combinations of these ingredients can produce a final product which is interesting enough to "elbow the salt out of the tongue's way" and still produce a satisfying result.

Until a few years ago, for me **spices and herbs** meant whatever came in a big plastic shaker from the grocery store.  There are many good options available that way, but now I take my time choosing spices and stock my spice cabinet with more finicky selections.  It is worth the effort to find the most flavorful versions of these little gems.  High quality garlic powder, onion powder, peppercorns, blends, Italian herbs, paprika, curry, cumin, garam masala, and others can brighten any dish.  My favorite store for these items is Penzey's™, a national franchise chain with stores in about half the United States, including one location in Houston.  They also maintain a very good web site (*penzeys.com*) for people who do not live near a store location.  The flavors of their products are just more intense, and this greatly assists in turning out better low sodium meals. When visiting one of their stores, virtually everything is available for sniffing, and items can be purchased in different sizes, with larger quantities resulting in substantially lower prices. I go through a large volume of spices and herbs, so Penzey's™ is my go-to store. I also like many of the items available at Whole Foods™, but even the regular grocery stores have good items that are generally not available elsewhere. For example, I continue to use Tony Chachere's™ Creole seasoning, despite having 310 mgs. of sodium for a quarter teaspoon, because it is so good! It is on my "do not disturb" list. This is one of the worst violations of my own general rule about sodium use, but I do so with malice aforethought. Let me explain how I get away with this felonious a-salt on my system. A very satisfying no-salt alternative to this product is Chef Paul Prudhomme's™ Magic Salt-Free Seasoning. I "de-salinize" the Tony Chachere's™ seasoning by diluting it with the Prudhomme™ blend. I stick to a simple one-to-one ratio. The net result is great flavor and manageable levels of sodium. Hail to Louisiana!

These are the spices that do a great deal of the heavy lifting in my kitchen. From Penzey's™, the Smoked Spanish Paprika is heady and rich and can be added to almost any savory dish. My favorite uses are on egg dishes, any recipe that calls for paprika, to flavor pan cooked or steamed green vegetables, roasted potatoes, and almost any recipe that is Spanish, Tex-Mex, Southwest or Cajun. Penzey's™ cumin, onion powder, garlic powder and curry powders are all outstanding. They just seem to impart more flavor than many other brands. The chipotle powder is the best smelling spice in the whole store as far as I am concerned. But it is blazing hot, so use it sparingly unless you like to break out in a sweat over your bowl of chili. Speaking of chili, Penzey's™ has a full line of chili powders, and I really love the chili con carne powder as a go-to spice for chili.

Herbs add a great layer of flavor when combined with these spices. Again, Penzeys™ has some superior choices, like their rosemary, which is my favorite. Also Penzey's™ Italian seasoning mix, their dried basil, thyme, and tarragon are among my favorites. Whatever green leaf herb you like, consider experimenting with it in a low sodium recipe. I tend to add a larger quantity of herbs and spices than I would normally use in a fully salted recipe that calls for herbs and spices as well. I often use twice as much spices in the low sodium version. Even if it seems like your are using a lot of these seasonings, these ingredients cannot hurt you (other than the hot peppers!), and if they can cover your taste buds in complex flavors, then this is really quite a good thing.

**Citrus and sours** are very important to keep layering on the flavor. Lime zest and juice, and lemon zest and juice can replace the bite of salt on the tongue pretty effectively. Vinegars can accomplish the same thing and can be used in a variety of items besides the traditional salad dressing. The bread recipe near the beginning of this book contained apple cider vinegar, and just does not taste as good without it. Most people enjoy salad dressing made with either red wine vinegar or a dark balsamic vinegar. Combined with olive oil and a good seasoning blend, like Mrs. Dash™ Garlic and Herbs, they create a very good basic salad dressing that can stand up to guests who normally consume salt in their dressing. We have found a variety of dark balsamic vinegars - traditional, cherry, raspberry, and others. They are all worth a try, and bring nice variations to a recipe. Perhaps lesser known is white balsamic vinegar, now widely available in regular grocery stores, which vinegar purists might tut-tut as not being true balsamic vinegar. Let them have their "tut-tut." White balsamic vinegar brings a more subtle taste to salads, very different in flavor from the dark varieties, and does not darken the color of the salad ingredients the way dark balsamic vinegar does.

White balsamic vinegar is a great item to add to a marinade for skinless boneless chicken breasts so that they grill up just a bit tangy and smoky without turning a dark brown from the traditional balsamic vinegar. Fish can also be marinated with white balsamic vinegar prior to grilling or pan cooking. Lemon, lime and white balsamic vinegar can be exchanged with each other to vary the taste of many recipes that call for either a citrus or a sour. New combinations can dance on your tongue and keep your focus away from the sharp zing of what was once the salty version of whatever you are cooking.

The final layer of flavor would be **heat and smoke**. Although not suited for every recipe, when either of these flavors is called for, the enhancement of a recipe can be exhilarating. A dash of Tobasco™ can really perk up an unsalted version of eggs, potatoes, vegetables and other dishes. A dash would contain very little sodium. Hot peppers of all kinds, whether fresh, dried or ground into powders, can take your mind off salt, and replace it with thoughts of cooling waters. It is an effective attention grabber, crowding out the taste buds that would otherwise be clamoring for the little white crystals. For smoke flavor, a few dashes of Liquid Smoke™ are excellent, and most brands contain either no sodium at all or a negligible amount. Too much

Liquid Smoke™ can taste a bit fake, though, but a modest amount in burgers, stews, chilis, bean dishes and even in Asian recipes can be the final accompaniment to all of the previous layers of flavor described in this chapter. Do not be afraid to use them all in the same recipe, or at least a combination of several flavor layers. But one word of caution about smoke – many packaged food products that come in a smoked version are usually high in sodium. Just because smoke flavor can enhance a dish that is otherwise low in sodium, you must still read labels carefully to ensure that your favorite smoked item isn't doused in sodium as well. You can be sure that smoked meats and cheeses are not on the low sodium list!

# VI.
# Workhorse ingredients

The ingredients mentioned in the *Spices of Life* chapter should all be considered heavy duty staples in a low sodium pantry. When used in abundance they will play a major role in making bland low sodium dishes taste much more palatable. Accordingly, every ingredient in the preceding chapter can be considered a "workhorse ingredient." In my vocabulary, workhorses are those items that you turn to over and over again for enhancing the taste of low sodium meals. They may also provide health benefits and can be used broadly in many types of dishes. Additionally, the spices and flavorings mentioned in the *Spices of Life* chapter do not add any meaningful calories, fat, sugar, cholesterol or carbohydrates, and yet many have health benefits such as antioxidant properties, promote heart health, aid in lowering blood pressure, and some can aid digestion. The hottest of these items can even stoke up the metabolism.

Although every ingredient in the *Spices of Life* chapter is a workhorse, a few stand out as the most helpful or versatile. I turn to them constantly, and could not sustain the 1,500 mgs. of daily sodium intake without access to these ingredients. So for my taste buds - and this is a personal preference by individual - the best of the best would be smoked paprika, rosemary, cumin, Cajun seasoning blends, chipotle powder, chili con carne powder, white balsamic vinegar and Liquid Smoke™.

The *Spices of Life* are not the only workhorses in the low sodium pantry. There are a few more that work well for me and should be given an earnest try by any reader of this book.

For example, Pacific™ brand Low Sodium Chicken Broth is a great base product for making many dishes. At only 70 mgs. per one cup serving, it is among the lowest sodium products of its type available. It can be found in many grocery stores in the soup aisle, where the aseptic containers of soup stock and broth are located. Substituting this broth for water, at least in part, can pump up the flavor of rice, pasta, beans, soups, stews, casseroles and countless other dishes. It adds a savory background without overpowering the other ingredients. There are several other brands, including some store brands, that also provide only 70 mgs. per one cup serving, so check all brands carefully. There are plenty of chicken broth choices that are higher than 70 mgs, all the way up into the many hundreds of milligrams per one cup serving.

Toasted sesame oil, which comes in many brands, is widely available in the Asian food aisle or even just in the cooking oil section of most grocery stores. This is a powerfully flavored oil; so it is used sparingly, primarily in Asian cooking. When combined with liquid smoke, garlic, ginger and other spices, the resulting Asian barbeque flavor is rich and makes added salt seem superfluous. Untoasted sesame oil performs none of this magic, so make sure you get the darker, toasted oil.

Simply Jif™ is a great tasting alternative to traditional peanut butter products. Unless you prefer the flavor of all natural ground peanuts, Simply Jif™ is smooth, creamy, slightly sweet and slightly salty, with only 32.5 mgs. of sodium and 1 gram of sugar per tablespoon. This is a good investment of salt and sugar, given the taste of the resulting product. This peanut butter is on my "do not disturb" list, and would be the peanut butter of choice even if sodium was not an issue. It can be used in many different types of snacks and is wonderful in a Thai inspired peanut sauce.

Whole Foods™ (and some other brands) blue corn and yellow corn tortilla chips have all the crunch and flavor properties of traditional chips but with a miniscule amount of sodium, in the range of 40 - 50 mgs. per 1 ounce serving. While there are a handful of no sodium snack items available, I personally cannot get too interested in the flavor of these ultra plain snacks. In my opinion, the small amount of salt used to make this lightly salted version is well worth the expenditure of sodium. They are great plain or in combination with many types of items to make snacking healthy and tasty. They are made from 100% corn, and do have a modest amount of fiber. They are also a good accompaniment to other casual low sodium dishes, rounding out the plate, so to speak. The Whole Foods™ version is modest in fat and low in sugar, adding to the overall attraction. You won't miss the 120 - 300 mg of sodium in chips once you keep these in your pantry.

The changes described thus far, and the food items introduced to this point, are the backdrop for a different way of shopping, cooking and combining of ingredients. The remainder of this book will focus on different types of cuisine; whether it be Asian, Mexican or Italian, or different type of foods such as soups, snacks and bean dishes. One chapter is devoted to Thanksgiving, the favorite holiday of so many, evoking all sorts of warm and homey images, but also one of the most treacherous meals on the entire calendar. A 500 mg. sodium Thanksgiving dinner is not only possible, it is actually delicious if you make the effort! The principles that work for Thanksgiving would also work for Christmas and other holidays on your personal calendar. This book finishes with dessert, just like any good meal itself. It is relatively easy to turn out good low sodium desserts. There will be tips and ideas tucked into the pages here and there. Everything is designed to help you succeed in turning a high sodium lifestyle into a healthier low sodium journey through this salted land of ours.

# VII.
# Cool Beans

Some of the most ordinary ingredients have found their way into the center of my low sodium world. Beans are a great example of this. They are very healthy complex carbohydrates that are necessary for sustained energy. They contain fiber to make you feel fuller and to fight cholesterol, have a solid amount of protein, contain almost no fat, and have essentially no sodium if you make the right choices, explanation to follow. I rarely sit down to eat just beans by themselves, but use them in combination with many other things to produce foundational dishes in my journey through the salted land.

Meats, beans, vegetables and seasonings can be combined with nearly endless variety. These one pot meals can be healthy, hearty, tasty and low sodium all at the same time. Not a bowl of glop, but great ingredients producing great casual food. Everyday meals. Breakfast (for some people), lunch and dinner.

I have a method for making these types of dishes that minimizes the time needed per meal and maximizes the value and convenience of the finished product. I always make a large pot of whatever it is I am making, season the pot (explained below) and then transfer the finished dish into multiple meal size storage containers. I typically freeze about one-half of the containers, and plan to consume the rest over the course of the following several days. I typically make between 8 to 10 servings at a time. This greatly cuts down on food preparation time during my busy work week. There is nothing new or unusual about these techniques, but they are especially important on a low sodium diet because if you don't have the time each day to prepare a thoughtful low sodium meal, you will typically end up reaching for what is easiest but not necessarily low in sodium. It's an easy pattern of behavior to fall into when busy and stressed. It can sabotage your effort to stay within a low sodium meal plan. I avoid this pitfall as much as possible by taking the time during the weekends to make larger batches of whatever I am preparing. Many of the chilis, soups and stews need plenty of simmer time, so I can be multi-tasking while the pot is on the stove on a low flame.

Over a few weeks time, having a stockpile of frozen meals ready to go, already divided and stored in proper portions, makes long term sustainable success just that much more likely.

Chili is a great place to start. A good chili can contain just about anything you like. I follow the principle of a well rounded meal in a pot, so I don't make chili from just meat, for example. The beans are a major co-star in this dish. My favorite beans for chili are red beans, dark red kidney beans, pinto beans and black beans. They are good by themselves or in combination. They are easily available in the dry form, typically in a bag on the same aisle as rice. Plain dry beans simply present no sodium issues whatsoever. They either contain zero sodium, or a truly negligible amount such as 5 to 10 mgs. per serving. Be wary of beans that are packaged as chili or soup in a bag, since many of these products typically are pre-seasoned and are likely to contain a high amount of sodium. As always, read every label. If a bag of beans does not come with a nutrition label, don't buy it! It is the only way to be sure of what you are getting. The Frontier Soups™ brand of dry soup and chili mixes is actually pretty good at packaging a variety of very low sodium mixes with great flavor, although not every item they sell is low in salt. You have to modify the cooking instructions that come with each of the low sodium dry mixes, but at least the starting point for the low sodium dry mix is well seasoned and not pre-packaged with a high amount of salt. Frontier Soups™ maintains an excellent website (*frontiersoups.com*) where all of their products can be purchased on-line, and they have an extensive product line.

With canned beans, you also have to pay very close attention. I have seen canned beans range from 10 to 500 mgs. per serving! Unless a can states "no sodium added," "no salt added" or "very low sodium," it is likely to have way too much sodium to be helpful in a low sodium diet. Most grocery stores stock "no salt added" canned beans in the same shelf area as the regular beans. A few stores also have a specialty area, like a health food section, with low sodium beans and other products. Whole Foods™ has a special section of no sodium added canned beans in most of their stores. The beans on these shelves range from 10 to 30 mgs. per serving, and are all acceptable for this purpose. I only infrequently see red beans in the low sodium canned form. Eden Foods™ makes a good no salt added canned red beans. This is another company with a good website (*edenfoods.com*) that contains nutrition information and where you can buy products online. The other bean varieties mentioned above are usually available in cans. There are numerous beans that some people may like in chili or other dishes, such as white beans and multi-colored varieties. So, experiment with all of them. Kuner's™ brand of Colorado (*faribaultfoods.com*) makes a good variety of no salt added beans. They also make a no salt added chili bean with seasonings already in the can, which is a good start on the way to making chili. Overall, dry beans are a much better bargain than canned beans from a cost standpoint, and take up far less storage space as well. The downside is that dry beans require a good long soak, which I typically accomplish overnight, plus boiling for at least one hour before placing them into a pot of chili for further cooking. So a little preparation and planning will save money and guarantee a no sodium bean for your dishes. I also help the beans along in the flavor department by adding some of the Chef Paul Prudhomme's™ Magic Salt-free Seasoning to the water when boiling. By itself, it does a great job of putting some background flavor into the otherwise very bland beans. Combining other spices and herbs is a great way to experiment and develop your own secret recipes.

I have used the following main protein sources for chili – ground beef, diced stew meat, diced round roast, diced turkey breast, ground turkey and diced or shredded chicken breast. Each imparts its own flavor to the chili, and the varied textures are also something to keep in mind. The sodium content of these proteins falls into a fairly narrow range. The chicken breast or breast tenderloin from Costco™ has only 40 to 50 mgs. per 4 ounce serving. The turkey breast products are around 75 mgs. per 4 ounce serving; ground beef contains around 85 mgs. per 4 ounce serving, and the stew meat will vary according to the package label. I have been fortunate to find numerous stores that carry stew meat around 85 mgs. per 4 ounce serving. The round roast is the wild card. If you cannot find it with a nutrition label, assume it is similar to ground beef. But it could be lower or higher than that, so I limit my purchases to those with a label. I always want to know, because I am counting every single milligram.

If the meat you typically buy has higher levels of sodium than the ones  mentioned here, you can definitely do better in most areas of the country. Chicken breast, even so-called "all natural" chicken, can have added sodium that takes a 4 ounce serving up to 300 to 500 mgs.. In addition to Costco™, Tyson™ and Sanderson Farms™ are examples of two brands available at many stores that have the low sodium package available. Besides sodium, some meats are much better than others in the fat department. I only use 96% lean ground beef or leaner. If fat is a serious concern, at least in the Houston area you can buy 98% lean ground beef made by Springer Hill™, available at Randall's™ grocery stores. I've used it for chili and it is fine. There may be a similar product in other areas of the country, so ask your grocer. I prefer the 96% lean, since the slightly higher level of fat does improve taste to a modest extent. I only use ground turkey that has 1 to 2 grams of fat per 4 ounce serving. When it comes to beef or poultry, I would much rather add fat from olive oil and get the benefits of using non-saturated fat than having considerably more animal fat in the pot. People who like dark meat chicken, turkey or duck in their chili will probably still wish to use it, assuming they can confirm the sodium content. It will provide a much higher fat content than white meat. Although I have never ventured into using venison, elk, rattlesnake or other more exotic meats, there is no reason why these products have to be laden with salt and they can be excellent choices as well.

Vegetables in chili? Now we've moved into the sometimes uncomfortable zone of personal preferences, family traditions, perceptions, myths and legends. I always add hefty quantities of the following finely diced vegetables to my chilis – onion, red and/or green bell pepper (or other colors on sale!), celery and carrots. I have used fresh, frozen and dried bell peppers, and they all work fine. On many occasions, I might also add one or more of the following: corn, peas, lentils, diced string beans, shallots, scallions and okra. I intend for chili to provide all of the benefits that vegetables have to offer – fiber, vitamins and minerals, antioxidants, low sugar, varied textures and  layers of flavor and aroma. Corn is somewhat of the exception, being higher in calories and sugar than the green vegetables. Peas and lentils are legumes and are closer in properties to the beans, so they can give chili a further protein and fiber kick. Skeptics who have tried my chili concoctions have uniformly indicated that the vegetables were a good addition. Maybe they were just being polite? But there is no reason not to try these one pot solutions so that you can more readily achieve a sustainable plan for healthy low sodium eating. It's just easier to make everything all at once.

*Season the pot.* This little phrase has been a great help to me. I start by using a large 5 quart Dutch oven, and determine the number of servings I am making. For chili, the standard is eight servings. If I am aiming for 300 mgs. of sodium per serving, then I have an allowance of 2,400 mgs. of sodium (300 x 8) that I can place in the pot. When I consider the use of salt in this fashion, it just seems to make everything easier to figure. Here is an example of how I use this "sodium bank:"

Meat: one pound of diced Costco™ chicken breast – 200 mgs.
Two 15 ounce cans low sodium diced tomatoes @ 20 mgs. per serving – 140 mgs.
½ small can of tomato paste – 30 mgs.
Celery – 2 cups chopped – 160 mgs.
Beans – two 14.5 ounce cans low sodium pinto beans @ 10 mgs. per serving - 70 mgs.
All other vegetables - 2 cups each chopped onions and bell peppers, plus 1 cup chopped carrots – 110 mgs.
½ teaspoon Tony Chacere's™ Creole Seasoning – 620 mgs.
Salt – ¼ teaspoon - 590 mgs.
Liquid Smoke™ to taste – nil
All other seasonings – nil (cumin, garlic, Rosemary and other Italian herbs, chili powders, black pepper, paprika, etc.)

This recipe would produce a pot of chili with just over 1,900 mgs. of sodium. Despite my 300 mgs. per serving allowance, this chili would come in at only 240 mgs. per serving. To raise the amount, I might use one pound of beef instead of chicken, or perhaps a bit more of the Creole seasoning, or even more salt. If I needed to lower the amount, I might switch to 1 cup of dry beans and save the 70 mgs. from the canned beans. The end result is a bowl of healthy, tasty chili with a very manageable amount of sodium.

Here is the rest of the recipe for this chili using the ingredients above without any variation in the sodium:

Brown the meat in 1½ tablespoons olive oil, and add onion powder, garlic powder, cumin, chili seasoning of your choice, Italian herbs, rosemary, black pepper and chipotle powder to taste. (You will be adding more of these same ingredients once the beans and vegetables are in the pot). Once browned, remove meat from the pot and add the vegetables (but not the tomatoes). Brown the vegetables first, using the retained oil and spices. Once browned, add beans, tomatoes, tomato paste and more seasonings, including any salt. Add back the meat and let pot simmer for one to two hours, stirring occasionally. Chili can be eaten immediately, or once cooled, transfer to serving containers for storage. The pot makes 8 meal-sized servings.

**Nutrition Information:**
Serving Size: 1/8th of pot
Sodium: 245 mgs.
Calories: 259
Total Fat: 4 g
Saturated Fat: 0.5 g
Cholesterol: 32 mgs.
Carbohydrates: 34.3 g
Fiber: 10.3 g
Sugar: 10.3 g (note: half of the sugar comes from the tomato products)
Protein: 20 g

From this base recipe, try the many variations of meats, beans, vegetables and seasonings to create your own favorite chili recipes. This is one of the healthiest dishes that you can prepare as long as a lean meat is selected. For example, you can spike up the protein with more lean meat. Add more beans for even more fiber and protein.

Ranchero beans are a close cousin to chili made with beans. They are a great bean and vegetable combination, and can be seasoned in varying degrees from very mild to multiple alarms of fire. They are a vegetarian side dish, made without a meat source, so they can accompany almost anything that is being prepared for a main dish. However, they are traditionally pared with barbeque, Tex-Mex or other Southwest fare. Any kind of bean can be used, but I use pintos most of the time and occasionally, black beans.

When making a side dish, the "sodium bank" for seasoning the pot has to vary from a main dish. I typically aim for 100 to 150 mgs. per side dish serving, leaving room for the other components of the meal. For 8 servings of ranchero beans at 150 mgs. per serving, that would give me a 1,200 mgs. sodium bank for the pot. Here is how I would use that bank:

Two 15 ounce cans low sodium diced tomatoes at 20 mgs. per serving – 140 mgs.

Celery – 2 cups chopped – 160 mgs.

Beans – two 14.5 ounce cans pinto beans drained of liquid at 10 mgs. per serving - 70 mgs. (Or make your own zero sodium beans by following the advice in the chili recipe above.)

Onions (yellow, white or red) – 2 cups chopped – 13 mgs.

Green, red and yellow bell peppers – 2 cups chopped – 10 mgs.

Cilantro – ⅔ cup chopped – 5 mgs.

Optional - corn or any other vegetables you like – nil to 50 mgs.

½ teaspoon Tony Chacere's™ Creole Seasoning – 620 mgs.

Low Sodium V-8™ juice – 1 cup - 140 mgs.

Dash chipotle chili powder (add more if you like it screaming hot!)

Ancho chili powder, all purpose chili seasoning, smoked paprika - to taste (I add lots of these ingredients, all of them with zero sodium)

Garlic and onion powder, all other seasonings – nil

Liquid Smoke™ to taste – nil

Optional – Tabasco™ sauce to taste – 28 mgs. per teaspoon

Olive oil – 2 TBSP – nil


In large pot or Dutch oven, heat oil and add all vegetables except for the tomatoes and cilantro. Add seasonings to the pot.

Brown the vegetables before adding beans, tomatoes and V-8™ Juice

Once everything is in the pot except the cilantro, let simmer for 30 - 45 minutes, gently stirring occasionally.

Add cilantro right before serving to keep it as fresh as possible.

**Nutrition Information:**
Serving Size: 1/8th of pot
Sodium: 131 mgs.
Calories: 178

Total Fat: 3.5 g
Saturated Fat: 0.3 g
Cholesterol: 0 mgs.
Carbohydrates: 29.1 g
Fiber: 9.4 g
Sugar: 5.8 g
Protein: 6.5 g

This recipe can be modified in so many ways. The smoky flavor could be omitted, or scallions and shallots can be added to, or replace, the onions. The heat level is certainly a variable. Black beans make it taste earthier, while other types of beans would produce more subtle changes. But following the basic principal of seasoning the pot to the number of milligrams per serving that you require is how to make this dish part of your long term commitment to low sodium eating.

A final bean recipe to highlight their versatility would be red beans and rice. This is the classic Louisiana dish that is traditionally served with sausage, chicken, vegetables, seasoning and a whole lot of salt. The low sodium version below uses turkey sausage links and a low fat, low sodium salami that may not be readily available in all parts of the country. The pot contains around 2,600 mgs. of sodium and makes about 9 full meal servings.

## Red Beans and Rice (300 mgs. sodium per serving):
## Ingredients:

6 low sodium Jimmy Dean™ turkey sausage links – sliced into small pieces -- 980 mgs.

4 oz AHI™ brand low sodium salami – sliced into small pieces -- 380 mgs.

Note – this item may be difficult to find.  If so, omit and add 2 more sausage links

1 lb chicken tenders – sliced into small pieces -- 200 mgs.

1 cup dry red beans

2 cans no salt added diced tomatoes --- 70 mgs.

3 cups celery -- 240 mgs.

2 cups bell pepper -- 10 mgs.

2 cups onions - 13 mgs.

1 cup brown jasmine rice

½ teaspoon of Tony Chachere's™  Creole seasoning  -- 620 mgs.

⅔ cup of low sodium V-8™ Juice -- 100 mgs.

Olive oil – 2 Tablespoons

Liquid Smoke™ – 1 Tablespoon

Spices - smoked paprika, garlic, cumin, ancho pepper, or chipotle pepper, Chef Paul Prudhomme's™ Magic Salt-free Seasoning

## Directions:

Prepare dry red beans – use the overnight soak method and boil for 1 hour, drain, and then add to the pot

Brown all of the meats together with some of the olive oil and seasonings in a 5 quart or larger Dutch oven.  Remove meats from the pot when done and set aside

Brown all of the vegetables in the Dutch oven with some of the seasonings and olive oil

Add tomatoes, beans and meats and simmer for at least 1 hour on low heat.

Prepare rice one of two ways:
- Cook in separate pot of unsalted water and serve the rice on the bottom of the bowl or plate. If using white rice instead of brown rice, a rice steamer can be used instead of boiling rice on the stove; or

- Add dry rice grains to the bean, meat and vegetable combination. Cook for 2 minutes, stirring frequently. Add water sufficient to cook the rice, and cover the pot to simmer on low heat until all of the liquid is absorbed.

Makes about 9 servings

**Nutrition Information:**
Serving Size: 1/9th of pot
Sodium: 308 mgs.
Calories: 339
Total Fat: 7.7 g
Saturated Fat: 1.8 g
Cholesterol: 49 mgs.
Carbohydrates: 40.3 g
Fiber: 6.9 g
Sugar: 7.2 g
Protein: 25.6 g

This dish is full of the flavors of Louisiana. As prepared here, it has a good deal less calories, fat, and of course, sodium, than the traditional preparation, while still retaining a healthy amount of fiber and good overall nutrition. An even lower sodium version of red beans and rice still tastes very good and can be accomplished by either reducing or eliminating some or all of the sausage, salami, Tony Chachere's™ Creole seasonings or the V-8™ juice. Add more chicken to replace any lost protein, and add more of the Chef Paul Prudhomme's™ Magic Salt-free Seasoning or other spices to replace the other reduced or eliminated ingredients. This dish would still have good flavor at 200 mgs. of sodium per serving or even less. But at 300 mgs. of sodium, you can probably serve this to unsuspecting friends and relatives and no one will be heard to say that your dish tastes like air, a favorite tease line from my older daughter! Some of your guests might still reach for the salt shaker, but you may be able to completely fool some of them, and they just might consume your fabulous concoction as is.

# VIII.
# In the Soup

Just like with chili, there are endless possibilities and varieties of soups. Most restaurant soup and store bought soups are extremely high in sodium. Since soups are mostly liquid, the added salt flavors the large quantity of the mostly bland liquid. I simply cannot order soup in a restaurant, and that is a small deprivation that I can live with quite comfortably.

I have sampled all of the available very low sodium soup products on the grocery store shelves, and frankly, I cannot find one that does not require lots of modification in order to make it palatable to my taste buds.

Creating a good very low sodium soup requires overcoming the challenge of flavoring the bland liquid base of the soup with ingredients other than salt. From my experience, the thicker the soup base, the easier it is to successfully flavor the soup without salt. So I tend to use dry split peas, lentils, rice or barley – either alone or in combination – as the base for most of the soups I make along with water and the low sodium chicken broth. They are just naturally thick, rich and hearty, and seasoning up from that starting point is not as difficult a task. These base ingredients are found dry in plastic bags or boxes, typically in the rice aisle of most grocery stores. Canned versions, if you can find them, are likely to be high in sodium; so unlike beans the dry version is the only realistic option. But there is an upside. These ingredients do not require an oversight soak, and do not need boiling prior to use in the soup, although you can certainly pre-boil any of them if you prefer. But they will do all of their softening and expanding inside the soup, which is the main reason for using them in the first place. Just like dry beans, they are very economical, and the peas and lentils pack a big nutritional punch, as they are quite high in protein and fiber.

Flavor is enhanced in the soup by use of the same principle I described as in making chili – a one pot meal consisting of meat, beans, grains or legumes, vegetables and seasonings. In fact, there are only a few small changes made to turn some of the chili dishes into a hearty soup. Replace the beans with peas, rice and barely, and modify the seasonings to some extent, and the transformation from chili to soup takes place. This is clearly demonstrated with the soup that I like to make most often, a rich split pea soup with chicken and vegetables. I like this soup for many reasons. It tastes pretty close to the soup that Grandma used to make. Plus, it is very high in fiber and protein. It's real comfort food on a cool day.

# Hearty Split Pea Soup
## (250 mgs. sodium per serving)

**Ingredients:**
1 ¼ cups dried green split peas
½ cup dry green lentils
½ cup dry pearled barley
1½ - 2 cups frozen mixed vegetables – corns, string beans, carrots, onions, peppers, celery, etc. (50 mgs. per cup)
2 Tablespoons olive oil
4 cups Pacific™ low sodium chicken broth (70 mgs. per cup = 280 mgs.)
1 lb cooked skinless chicken breasts, diced into small pieces (200 mgs. per pound)
8 cups water
½ teaspoon salt – 1,180 mgs.
Seasonings to taste – garlic powder, onion powder, cumin, rosemary, paprika, Liquid Smoke™, etc.

**Preparation:**
Use a non-stick 5 quart Dutch oven to prepare the soup.
Brown chicken in olive oil and a small amount of the seasonings.
Add water and chicken broth to the pot and bring to boil.

Add all of the ingredients and reduce heat to simmer.
Let simmer for 2 hours on low heat, stirring occasionally.
Add more water if soup becomes too thick.
Once the peas have fully broken down and soup is thick, it
is ready to serve, or let it cool somewhat before transferring
into storage containers.
Will make 8 main meal servings.  Keeps in refrigerator about
1 week, and also freezes well.

**Nutrition Information:**
Serving Size: 1/8th of pot
Sodium:  240 mgs.
Calories: 320
Total Fat: 4.6 g
Saturated Fat: 0.5 g
Cholesterol:  32 mgs.
Carbohydrates:  43.3 g
Fiber:  15.4 g
Sugar:  3.1 g
Protein: 27.2 g

Variations in the protein sources, vegetables, base ingredients and seasonings can produce many different soup dishes.  Some might be thicker while others are less thick.  For an easy change, there are also many varieties of lentils available, especially in the bulk sections of stores like Whole Foods.  They range in size, color, flavor and some have very high levels of protein and fiber.  If using very high protein lentils, you can reduce the beef or poultry portion and still have a full complement of protein.  This reduction in meat can also result in less sodium going into the pot, or it can leave room for a small increase in salt.

Pea soup or the modified version which becomes a lentil soup can both be especially tasty if a modest amount of finely diced turkey bacon is added to the preparation. Each strip of turkey bacon will likely have around 180 to 200 mgs. of sodium, so 6 strips should replace the salt listed in the above 8 serving recipe. The smoky bacon flavor will enhance the overall flavor complexity of the soup. This process of substitution is a very important activity that each individual can experiment with, finding the right combinations that work for the particular palate. If it turns out that someone just loves turkey bacon, try using it in a variety of recipes in this book or recipes from other sources, and make any needed adjustment in other ingredients containing sodium. Remember the sodium bank concept as you experiment. Turkey bacon is low in fat and there are several brands that contain a good amount of protein per slice, so I greatly prefer it over traditional bacon, which does not fit into my nutritional goals. The Applegate Farms™ and Wellshire™ brands are examples of higher protein turkey bacon. They cost more than the mass market brands, but the higher protein is the reason to purchase them.

Continuing with some suggested modifications, a very good chicken and vegetable soup can be made from the above recipe by eliminating all but a small amount of the split peas and lentils, yet leaving all or some the barley. By adding 2 cans of no salt added tomatoes, the broth can be transformed from a thick pea soup type of base to a thinner tomato-based stock. Diced potatoes or small cut pasta could be added for heartiness without adding any salt. This could result in a minestrone style soup, one of my all time favorites. A small amount of any type of beans could also be added. The Liquid Smoke™ might also be omitted in this conversion. If the soup base seems too thin or runny, a small amount of tomato paste can add some thickness, and it is naturally very low in sodium. Of course, a beef version of this soup could also be created, and diced mushrooms are a good addition with beef. And so on, and so on.

Another approach to making a tasty low sodium soup is to modify a low sodium store bought soup product. Campbell's™ Low Sodium Cream of Mushroom Soup, for example, is truly a low sodium product with only 60 mgs. per 1 full can serving. But in my opinion, it needs a little help! I take 2 cans (120 mgs.) of this soup, add 2 cups of Pacific™ low sodium broth (140 mgs.), tons of my favorite seasonings, plus my favorite vegetables, and make a soup with about 100 mgs. sodium to accompany a main dish. By adding a protein dish, a full meal has been created with sufficient protein. Prior to modifying, the soup only contained 4 grams of protein per 1 can serving. There is even less protein per serving in the modified version, but the main protein dish takes care of this need anyway. This recipe makes enough soup for 4 fairly large servings, and as a side benefit, the fat content of the soup has been reduced from 8 grams in the original can, to around 4 grams per serving after the changes are made. Although Campbell's™ also makes a low sodium chicken broth, it is higher in sodium than the Pacific™ brand, so I don't use it.

This modifying technique might thin down a soup that started as a thicker product. If you prefer a thicker soup, you can add your own thickeners without increasing the sodium content – potatoes, rice, beans, barley, peas, and lentils all cook down to add a thickening element to the broth. It may take a while for these thickeners to do their job, so I recommend that you make more than just a couple servings of this soup at a time and freeze some so that soup making can be an occasional project when you have the time. You can always multi-task while your fabulous concoction is simmering on low heat, making it more likely that you'll take the time to make this soup in the first place. Another tip is to boil up these thickening ingredients at any time and keep them at the ready to add to soups as you make them. They will all freeze well once cooked.

There are a growing number of "lower sodium" soups on the market, and left unmodified, they can be a bit of a problem. Most of these products have taken an original 700 to 1000 mgs. soup product and reduced it to 350 to 600 mgs. Yes, they are lower in sodium and they taste pretty good too. However, they are not very workable in an over-all low sodium diet targeting 1,500 mgs. per day. In a pinch, turn one serving of tasty lower sodium soup into two servings of an even lower sodium soup by adding one can's worth of Pacific low sodium broth, and enough protein, vegeta-bles and seasonings to turn this into two full meals. Most of the Campbell's™ lower sodium soups contain between 410 - 450 mgs. of sodium, and only a moderate amount of protein. The Progresso™ line ranges from 400 - 480 mgs. of sodium. If your additions take the pot up to 600 mgs., but result in two complete meals including protein, then the 300 mgs. bowl of "main meal" soup is not a bad compromise.

This technique opens up a wide range of possibili-ties as this type of soup product comes in many offerings by many different manufacturers. You have to experiment and develop the combinations that suit your own particular tastes. After you have accomplished this, and recorded the recipe for future use, then you will have added to your arsenal of weapons to combat the salted landscape in which we must make our way.

Beef stew is one of the classic comfort foods that can be made in one pot. From the descriptions of the chilis and soups in this chapter and the prior one, it is pretty easy to see how a low sodium version of beef stew would come together. Here is my favorite recipe, which makes 8 meal size servings, each under 300 mgs. of sodium per serving:

# Beef Stew

**Ingredients:**
2 pounds stew meat at about 75 mgs. per 4 ounce serving – 800 mgs.
2 pounds of any potatoes, cut into bit sized pieces – 100 mgs.
1 pound carrots, cut into chunks – 300 mgs.
1 pound pearl onions, or 1 pound yellow onions diced large – 40 mgs.
2 cups diced celery – 160 mgs.
2 - 15 ounce cans no salt added diced tomatoes – 140 mgs.
I cup Pacific™ brand Chicken Broth – 70 mgs.
Rosemary, Italian herbs, garlic powder, black pepper, dried basil, thyme, bay leaf and other seasonings to taste – nil
Salt – ¼ teaspoon – 590 mgs.
2 Tablespoons olive oil
Optional – red wine to taste

## Directions:

Brown the meat in olive oil and the spices and herbs (but not the salt).

Add all of the vegetables except the tomatoes to the meat and cook for 5 to 10 minutes before adding liquids.

Add tomatoes, chicken broth, salt and optional red wine and use liquids to deglaze the bottom of the pot.

Reduce heat, cover and simmer for at least 1 hour, stirring occasionally. The liquids should thicken over time as the potatoes break down and the tomato juices condense. For a thicker base, add a small amount of tomato paste.

## Nutrition Information (without red wine):

Serving Size: 1/8th of pot
Sodium: 283 mgs.
Calories: 342
Total Fat: 8.1 g
Saturated Fat: 1.8 g
Cholesterol: 48 mgs.
Carbohydrates: 40.5 g
Fiber: 7.3 g
Sugar: 13.3 g
Protein: 26.5 g

This recipe may look like one that many people already use, except that in the traditional recipe there is much more sodium in the regular version of the canned tomatoes and more salt is typically added than just the ¼ teaspoon for this 8 serving recipe. These two changes alone make a huge difference in the level of sodium! The rosemary, tomatoes and other seasonings are the workhorse ingredients that help give this dish a savory, hearty flavor without the traditional amount of sodium. This one pot meal packs plenty of protein, modest calories, fat and sugar. It's a winner, especially on a cold night.

# IX.
# Snack Attack

Who doesn't like snacks between meals?  Starting at childhood, most of us remember snacks during the school day, after school snacks, or making our own from the pantry whenever we were hungry.  There are multi-billion dollar corporations who produce nothing but sweet snacks, salty snacks, crunchy snacks, sweet and salty combination snacks, and so forth.  What a way to blow a load of sodium! And who hasn't also been told at some point in their lives that snacking between meals is bad for you, that it adds pounds and can ruin your appetite?

So which is it – to snack or not to snack?  Experts may differ, not to mention moms.  Here is my answer -- the sustainable low sodium diet includes intakes of small amounts of sodium throughout the day, and snacking helps to ward off salt cravings and binge eating, when careful food selection is usually tossed out the window.  So I opt for snacks.

But I want to make good use of them. There is some planning that goes into making good snack choices. There is a simple framework that helps a great deal with planning the day from a nutritional standpoint. I usually plan for 3 regular meals and 2 or 3 snacks per day. This keeps the body from getting too hungry, and keeps fueling the activities as the day goes on. Meals should be smaller if snack meals are included in a daily plan. My typical distribution of sodium would be around 400 mgs. per regular meal, with about 150 mgs. per snack. Of course, each meal or snack can vary from this average, as long as the day finishes near the 1,500 mg mark. Building meals around these sodium guideposts can be challenging at first, but over time it becomes second nature. There are recipes throughout this book that fall well within these parameters. There are also many snack choices that fit this approach to meal planning. Many of these choices are fairly high in protein and fiber, and can help you get through the day, or at least until your next meal. Here are just a few examples:

- 8 ounce glass of milk, with a serving of fruit and ¼ cup of unsalted nuts -- 125 mgs. of sodium.
- 8 ounces of milk with 365™ brand chocolate chip cookies – these are small but tasty treats from Whole Foods™ which contain 18 mgs. of sodium each, and just over 1 gram of fat per cookie. The 365™ brand also has a

similar oatmeal cookie. Any similar cookie or cracker can be substituted.

- ½ cup of plain nonfat Greek yogurt, mixed with 1 tablespoon of jelly, preserves or fruit spread and chopped walnuts (or any other preferred nut) -- 65 mgs. of sodium. Add a few of the 365™ brand chocolate chip cookies – a serving of 5 would add 90 mgs. and bring the total of this filling snack to 155 mgs. with 6 added fat grams from the cookies.
- ½ cup of low sodium cottage cheese (Friendship™ and Lucerne™ brands are each around 50 mgs. of sodium per serving) with 1 ounce of hummus (choose hummus with about 75 mgs per 1 ounce serving), or 1 tablespoon of salsa (about 50 mgs or less sodium per tablespoon), and a small serving of low sodium tortilla chips (30 mgs.) – for a total of 130 to 160 mgs.
- ½ cup of low sodium cottage cheese with 1 ounce of hummus plus carrot sticks, celery sticks, bell pepper slices, cherry tomatoes – for a total around 150 mgs. of sodium.
- ½ cup of low sodium cottage cheese with 1 tablespoon of salsa (about 50 mgs or less per tablespoon) and 2 Snyder's of Hanover's™ Old Tyme Pretzels – for a total around 180 mgs. of sodium.
- ½ cup of low sodium cottage cheese and fresh fruit (peach, apple, orange, grapes, banana, cantaloupe) on an optional small bed of iceberg lettuce. Add ½ ounce dry roasted unsalted peanuts on the side. The whole mini-meal comes in around 60 to 70 mgs.
- 1 slice of toasted no sodium bread, 1 tablespoon of Simply Jif™ peanut butter (32.5 mgs. of sodium per tablespoon), plus your choice of fruit. This snack can be extremely low in sodium if using zero sodium bread, around 35 mgs. This is filling and delicious, and can help "catch up" the day if a meal on that day is higher in sodium than normal. Instead of fruit, an 8 ounce glass of skim or low fat milk kicks up the protein even further,

and still keeps this snack in the range of 150 to 160 mgs. of sodium.

- A slightly higher sodium version of the above snack would use 5 Hint of Salt Ritz™ crackers with the peanut butter, which only adds 35 mgs. of sodium, bringing the total to 65 to 70 mgs. Adding fruit would not change this total, but adding milk could bring this snack close to 200 mgs. of sodium, which is fine as long as the rest of the meals on that day are planned to accommodate this additional sodium.

- 1 stalk of celery filled with one tablespoon of light cream cheese -- 100 mgs. of sodium. To this, a 1 ounce serving of low sodium tortilla chips would add another 50 mgs. Or the 5 crackers plus peanut butter would add about 65 mgs. instead.

- A half portion of a meal that I would normally have for lunch or dinner makes a great snack, and would come in around 100 to 150 mgs. These meals are discussed throughout this book.

These snacks have helped to keep me on track. In addition to being low in sodium, they provide the protein and carbohydrates needed for energy and for feeling satisfied until the next regular meal. They really work! These are just a few of the myriad combinations of snack foods that can produce a beneficial mini-meal. Keep track of what is consumed during the day, and better yet, plan the meals and snacks ahead of time to make sure that your meal plan is well thought out and can be executed with the items that you have on hand.

The pretzels that I mentioned above are an answer to a question that I posed earlier in the book. I did have to give up the very salty pretzels that I used to love, but what I have learned is that most foods come in a range of saltiness, and pretzels are a good example of this. A serving can range from almost no sodium to upwards of 400 mgs., believe it or not. The Synder's™ Old Tyme pretzels are not specifically labeled as a low sodium product, but they have less than half the sodium per ounce than their saltiest cousins. This is a case of where it is hard to distinguish the saltiness of the two types of pretzels based just on munching alone. Having an occasional pretzel with ice cream or a cottage cheese snack, or in some other way that you like, is just much better than not having one at all. This helps the mind adjust to the changes that have to be made, without the overwhelming sense of deprivation. This sodium range is present in many products, so don't automatically give up on a favorite item just because you have been eating a high sodium version. At least take the time to look for substitutes, some of which are not even marketed for a sodium restricted diet, and don't taste like it either!

I also have a favorite "all day snack" trick that allows me to put something in my mouth whenever I want, to tide me over between meals, or even between snacks! I keep a small container of almonds and raisins with me at work. In the morning before leaving for work, I literally count out 18 dry roasted almonds, and add about 1 tablespoon of raisins and place them into a very small container with a firm lid. I don't allow myself access to any more of these items, since it is easy to gobble up handfuls of them, which is not a good idea, at least for me. So I don't keep bags or boxes at work – just what I bring from home. This snack contains zero sodium, but has protein and carbohydrates that just seem to help even out the day. A few pieces taken a couple of times in the morning, followed by the same approach in the afternoon, is a little additional insurance against going on an unhealthy and usually high sodium binge. While there is no guarantee that this approach will ward off potato chip and donut cravings forever – it certainly helps!

One snack item made by Whole Foods™ should be given honorable mention. They have taken the traditional cheese puff and knocked it down to size, sodium-wise, of course. One cup of 365™ brand cheese puffs contains 70 mgs. of sodium and only 3 grams of fat. I usually make a snack out of 1 to 2 cups of them in a plastic bag, and start snacking on them at lunch, finishing by mid-afternoon. They have all the good cheesy taste but far, far less sodium than the major national brands. It is hard to feel too deprived when there are products on the market like this one. It also comes in the "cheese curls" variety.

When my grown children first saw me snacking this way, they questioned my commitment to healthier eating. It looks like I am just eating whatever I want, whenever I want to eat it. That's the beauty of planning the day's meals ahead of time. I explained to them what I was doing and how I went about doing it, and now they at least don't automatically assume that I have fallen off the wagon when they see my with a handful of chips or cheese puffs. It can make you feel pretty good when you are seemingly eating items that taste like mainstream snacks. You know the work you've done to prepare them appropriately and fit them within an overall nutritional plan. Once this becomes second nature, it feels like normalcy has been restored to your diet as compared to the earlier chaos of making the adjustments to a low sodium lifestyle.

# X.
# The Backyard Barbeque

Fire up the grill and find that checkered vinyl table-cloth!   Great low sodium meals can be made with the smoky flavor of a good barbeque.   Smoke is one of the great flavor enhancers that should be used to help offset the lack of sodium.   There is no reason not to enjoy grilled burgers, chicken, steak, chops, ribs, cutlets, fish, shrimp, scallops and other seafood, with many of the typical side dishes and accompaniments that make the barbeque a classic American meal.

Consider this very basic barbeque meal plan for starters:

*Grilled cheese burgers and fixings on buns*

*Coleslaw*

*Potato salad*

*Ranchero beans*

*Pickles*

Is this a 3,000 mg. sodium extravaganza, or a 500 mg. low sodium feast?  Well, it could be either one.  But in this chapter, the entirety of this meal will add up to 500 mgs. of sodium or less.  It takes a bit of effort, but the ideas discussed below are a method for solving the fundamental sodium problem that can be applied to any kind of meal.

Burgers -- the same lean ground beef that was used in Chapter VII to make chili is a great starter for the burgers.  At 85 mgs. for a 4 ounce serving, it fits well within the parameters of this 500 mg. meal.  Use ¼ cup of no salt bread crumbs or matzah meal, 1 egg or ¼ cup of egg substitute per 1 pound of meat used, and add all of your favorite spices and herbs (from Chapter V) to the meat and mix well.  Of course, finely diced onions or shallots are a great addition, and some people like mushrooms and other veggies finely chopped into their burgers.  Form the meat mixture into 4 patties for each 1 pound of meat used.  With the egg added to the meat, each burger will contribute about 100 mgs. of sodium to the meal.  Place the patty on a medium heat grill which has been sprayed with cooking spray made for grilling, or brush the grates with a little olive or peanut oil.

When the burger is nearly done, at your option, add a slice of the Lacey Swiss cheese discussed in Chapter IV to the meat to melt right before removing from the grill, or add the cheese to the sandwich roll instead. Use the zero sodium hamburger rolls that are made from the recipe in Chapter III. The buns can be quickly grilled for added flavor. Use your choice of low sodium mustard, ketchup, or horseradish sauce on the bun, also described in Chapter IV. A tasty low sodium barbeque sauce might be a good choice instead. The no sodium ketchup would obviously add zero mgs., while the horseradish sauce adds 15 mgs. per teaspoon, and mustard choices range from 0 to 50 mgs. per teaspoon. Heinz™ No Salt Added Ketchup and Plochman's™ Honey Mustard would add no sodium at all to this sandwich. Meyer's Elgin Smokehouse™ Barbecue Sauce is an ultra-low sodium but delicious sauce. It contains a mere 55 mgs. of sodium for a 2 tablespoon serving, yet you'd never know it! The website for ordering this sauce is *cuetopiatexas.com*. Otherwise, it is available in a number of grocery stores in Texas. There are other low sodium barbeque sauces dotting a landscape which is mostly full of high sodium products. Another example would be Bone Suckin'™ Barbeque Sauce (available at Whole Foods™ and other stores as well as online at *bonesuckin.com*) with great taste at around 50 mgs. per tablespoon. However, the Bone Suckin'™ Barbeque Sauce is very sweet and adds 4 grams of sugar for each tablespoon, so I use it sparingly, but I do use it. There is less sugar in the Meyer's Elgin Smoke-house™ line of sauces. And there are numerous other low sodium barbeque sauces throughout the country which contain less sugar, so check with your grocer if this is important to you. Also feel free to add lettuce, tomato, onions and no sodium bread and butter pickles from Chapter IV (or make you own as discussed below in this chapter.) The entire sandwich can range from 130 mgs. of sodium on the low end to about 280 mgs. on the high end if a tablespoon of deli mustard was used.

The condiments can absolutely make the difference in the amount of sodium in this sandwich, yet for most people, the sodium content of all of these ingredients is not even an afterthought! A classic Dijon mustard can contain up to 120 mgs. per teaspoon, so the fact that there are choices ranging from 0 to 50 mgs. per teaspoon makes it worth the effort to read every label. This same burger could easily come in at well over 1,000 mgs. of sodium and could even reach up to 2,000 mgs. without too much difficulty. While it may taste better to some people, the difference in taste is not worth the vast difference in the level of sodium, especially if you are trying to maintain a healthy low sodium lifestyle.

**Coleslaw** – I use a simple, three ingredient recipe to make a very satisfying coleslaw. It is best to prepare it a day in advance, and allow the ingredients to come together and slightly soften the cabbage. I take 1 bag of the tri color slaw (green cabbage, red cabbage and carrots), the juice of ½ of a lemon, and 1 tablespoon per serving of Hellman's™ Canola Low Fat Mayonnaise. This bag usually makes about 5 servings, so this would result in 5 tablespoons of mayonnaise, which contains 450 mgs. of sodium. It sounds like a lot of salt, but as I discussed in a prior chapter, this product is so good that the salt is a smart investment and turns this side dish into something that might come close to passing for the regular coleslaw preparation. I like to squeeze the juice of the lemon into the bottom of a large mixing bowl, and then add the mayonnaise. I mix the two together until the lemon juice has been fully incorporated and the mayonnaise is creamier as a result. Then toss in the bag of slaw and mix well. It is easier to get a good mix when the mayonnaise is a bit creamier. Transfer to a storage container and place it in the refrigerator for an overnight rest. It will be creamy and delicious the next day even though it may look pretty dry when you first mix everything together. A 1/5 bag serving of this product comes in at around 100 mgs. of sodium.

**Potato Salad** – Cut up 2 pounds of potatoes and either boil in water or roast in the oven until tender. Yukon gold potatoes and red potatoes are both good choices. There's no question about it – roasted potatoes make a more intensely flavored potato salad, but many people boil them in water anyway. Try it both ways and see if the low sodium version tastes better to your taste buds one way or the other. In any event, there is very little sodium in the potatoes. While the potatoes are still warm, transfer them into a large mixing bowl and add either diced yellow, white or red onion, scallions or shallots – your choice. The sodium is negligible, with a full cup chopped being less than 10 mgs. Add a full cup of chopped celery, which does contribute 80 mgs. per cup. Next use 5 to 6 tablespoons of the Canola mayonnaise, which adds up to 540 mgs. of sodium to the bowl. Add rosemary or herbs to taste if you like. Combine and chill, preferably overnight to allow all of the flavors to meld. This basic recipe would yield about 5 - 6 servings, and at 6 servings, the sodium per serving would be around 110 mgs. You could also garnish the potato salad with parsley and paprika just before serving. A mustard version can also be made for around the same amount of sodium, but you would have to choose your mustard carefully. A combination of mayonnaise and mustard might be ideal, so make sure you give these varying combinations a try.

**Ranchero Beans** -- follow the recipe in Chapter VII, and a serving of the beans would have about 130 mgs. of sodium.

**Pickles** – In addition to the zero sodium bread and butter pickles discussed in Chapter IV, I make my own less sweet low sodium pickles using Penzey's™ pickling spices, vinegar, sugar, garlic , onions, water and a small amount of salt. I like to cut 4 or 5 pickling cucumbers in half, and then slice each half lengthwise to produce spears. You can also just slice the cucumbers into round disks or on a bias for a different shape. The pickling cucumbers are often smaller than regular cucumbers, about 5 to 6 inches long. They are also thinner than regular cucumbers. They are typically seedless and the skin is very good to eat once pickled. However, any cucumber can be used to make pickles. Prepare about ½ cup of sliced onions to add to the jar at the same time that you place the pickles in the jar. To make the low sodium brine, place 1 cup of white vinegar into a small saucepan with 2 tablespoons of sugar and 1 cup of water. Add 1 heaping tablespoon of pickling spices, ¼ teaspoon of salt (which puts 590 mgs. of sodium in the brine), minced garlic, or 2 to 3 thinly sliced garlic cloves. Bring this mixture to a boil. Let the brine simmer for a minute or two, and then let it cool completely to room temperature. Place the sliced cucumbers into a vertical jar, and add the onions into the jar as well. Fresh dill or garlic cloves would make good additions. For 5 pickling cucumbers, I would recommend using a 32 ounce jar. Carefully pour the cooled brine and all of its contents into the jar and leave about a ½ inch of space at the top of the jar. If the liquid isn't enough to fill to this level, add water to get to the right level. Screw on the lid and store in the refrigerator for at least a couple of days before eating. I like to gently shake the contents a couple of times, but I have no idea if this does anything beneficial to the pickling process. It just feels like it's the right thing to do. A small amount of the salt will be absorbed into the pickles, but some of it will remain in the brine. A pickle spear made this way will contain a very small amount of sodium, around 10 mgs., but is big on flavor. They can last in the refrigerator for several weeks, but seriously, they won't be around that long if you like the taste of these healthy, low sodium pickles.

When it comes to loading up your plate at the bar-beque, most people will not typically take full servings of each and every one of the items on this menu, but assume that you have not eaten all day long and are starving, and decide to completely pig out anyway. Here is the sodium bank for this huge, casual meal:

Hamburger, cheese, ketchup, honey mustard, lettuce, tomato, pickle, bun
– 150 mgs.
Coleslaw – 100 mgs.
Potato salad – 110 mgs.
Ranchero beans – 130 mgs.
Pickle spears – 20 mgs.
Total – 520 mgs. of sodium.

Personally, I would be hard pressed to consume full serving portions of all these different dishes, so I would likely end up taking smaller portions of the side dishes, consuming about 400 to 450 mgs. of sodium for the entire meal, and still be one happy (and stuffed!) camper. So there it is – a great barbeque meal for under 500 mgs. of sodium.

Chicken breasts can be marinated in olive oil, white balsamic vinegar and spices for an hour and placed on the grill. No skin is needed to produce a very good layer of grilled flavor. I tend not to use any vinegar with steak, chops, ribs, fish or seafood, but do like to use olive oil and spices as a marinade before grilling. They all turn out very well. On any of these proteins, a sprinkle of paprika, ancho chili powder or some other red pepper product tends to produce a slight crust from the grilling process. All of these protein sources can be prepared with only the naturally occurring sodium in the meat, which ranges from about 75 to 120 mgs. (ribs) for a 4 ounce serving. My favorite fish to grill are halibut, salmon and Chilean sea bass, all with the skin on. I like to place the skin side directly on the grill, and cook the fish without turning it. It will even cook through on a fairly thick piece, but may be slightly less done in the center. I like it this way. If you plan to flip the fish over, place some non stick spray on the grates and after flipping the fish, leave the flesh side on the grill for a few minutes before trying to take the fish off the grill. It will usually hold up pretty well that way, forming a slight crust on the flesh side. Very little of the meat should end up on the grill.

Other traditional barbeque side dishes and accompaniments can also be made the low sodium way. Some have already been mentioned in earlier chapters of this book:

Low sodium tortilla chips – 50 mgs. per serving

Salsa – 25 to 50 mgs. per tablespoon  (homemade salsa could contain even less sodium)

French Fries – Ore Ida™ makes French fries and steak fries with very little sodium, and there are other widely available products between 0 to 50 mgs. per serving. I bake the French fries in the oven, and sprinkle olive oil and seasonings on them before roasting. Smoky paprika works very well with fries. With the zero sodium ketchup, this tasty treat is much lower in sodium than potato salad, and has only a small amount of healthy olive oil compared to traditional frying. Once again, care in reading the packages is a must. I can pull one bag out of the freezer section of the grocery store and it contains 330 mgs. of sodium per serving. With a slight change in the package description, the next bag is under 50 mgs. per serving. If the French fries are pre-seasoned, they will invariably be high in sodium, as are most curly fries, waffle fries, Southern style fries, and tater tots. So the key is to find plain potatoes in either thin strips or thicker wedges.

Hot Dogs and Baked Beans -- I have yet to find good substitutes for hot dogs and baked beans. They might exist in some parts of the country. The lowest sodium content I have found for hot dogs is still 290 mgs. of sodium per dog. To get the same amount of protein as in the hamburger, you would need to consume 2 hot dogs. This takes the sodium level to 580 mgs. for just the meat alone. This does not work for me. While I have tasted a commercially prepared can of low sodium baked beans, it was so different tasting than the traditional baked beans that it didn't work for me either. I tried to make my own recipe, but also failed to come close to the real thing. At least the Ranchero beans (from Chapter VII) taste like they are supposed to taste, and don't mess with my taste buds. But it is certainly possible that in various parts of the country, there are substitutes coming onto the market that are an improvement over what I have been able to find as of the time of writing this book. So, it is important to keep looking, keep asking, and keep experimenting — it will only get better over time.

# XI.
# Tex-Mex Fiesta

Tex-Mex cuisine just tastes like a party. Highly seasoned and highly flavorful, it has been growing in national popularity for years, and it is easy to see why. But it is traditionally a very salty cuisine, so modification is of major importance to make it part of the low sodium diet.

Several good Tex-Mex items have already been discussed in earlier chapters. Corn tortillas are the must-have tortilla to avoid loads of extra sodium, at only 10 mgs. per tortilla for the normal-sized versions available in most grocery stores. In restaurants, the corn tortillas may be larger, and may have more salt depending upon the whim of the particular restaurant, so making an inquiry ahead of time is critical.

Tortilla chips can be both good tasting and low in sodium, if you are willing to do the homework and find the brands in your area of the country. I have mentioned the Whole Foods™ 365 brand blue and yellow corn tortilla chips, and they are my go-to chips for Tex-Mex and general snacking. Garden of Eatin'™ also makes a very good blue corn chip with about 60 mgs. per serving. There are other brands throughout the country worth checking. A local Texas brand, H.E.B. Central Market™, makes a Hatch green chile chip that has 80 mgs. of sodium per serving and has knock-out green chile flavor with a bit of heat. I keep these on hand for occasional snacking, and it is fairly easy to stay within 1,500 mgs. per day even with this slightly higher sodium version of chips. I sometimes combine one half regular chips with one half hatch chips. I just do the math for the day and make sure my other choices fall within the parameters. Another really tasty chip with only 80 mgs. of sodium per 1 ounce serving is the Fritos™ Lightly Salted Corn Chips. They will bring back memories of snacking on the real thing. They are higher in fat than the other chips mentioned in the book, but they are an example of what manufacturers are doing to help slowly draw the populace off of its sodium addiction.

Salsa has been mentioned several times in this book. In addition to the peach salsa that is my favorite for most dishes, a zippy homemade salsa may work best for Tex-Mex dinners. Chopped tomatoes, onions, cilantro, garlic, sweet bell peppers, lime juice, chipotle and other hot peppers to taste, cumin, adobo and other favorite seasonings make a very good salsa without a drop of added sodium. Be creative with the assortment of vegetables and seasonings, and salsa can be produced in endless varieties to suit every taste and menu. If you prefer the thicker sauce-laden versions typically found in the bottle or jar, try spiking a small amount of low sodium V-8™ Juice with Mrs. Dash™ Fiesta Lime seasoning, a splash of Tabasco™ sauce, and add that to the chopped vegetables. I also like other fruit salsas besides peach using most of the same ingredients in a regular salsa but adding diced pineapple, mango, pears or anything else that pleases the chef. Corn and black beans make good additions to salsa, either by adding one or both of them to an existing salsa or by combining them to make a very hearty salsa that really stands up like a side dish.

Beans have been mentioned before as well. The Ranchero Bean recipe in Chapter VII works very well as a Tex-Mex side dish. Pinto beans and black beans tend to be the beans of choice for Tex-Mex cuisine. Chapter VII discussed bean preparation for dry beans as well as finding canned low sodium beans. Beans can be added to salsa, as just described above, served hot as a side dish, mashed and refried, or added as an accompaniment to fajitas or soft shell corn tacos.

Guacamole is made with the mostly bland tasting avocado, and a small amount of sodium may help the flavors along.  When I make it with no added salt at all, I use smoky ingredients like smoky paprika and cumin, plus hotter ingredients like peppers or chipotle powder.  Lots of fresh tomatoes and onions are a staple in guacamole.  I also like minced garlic, chopped cilantro and diced sweet bell peppers.  Lime zest and the juice from the fruit are also great complements to the flavor, as well as the Fiesta Lime seasoning made by Mrs. Dash™.  Some people prefer the blended version of guacamole (in the food processor, for example) while others, myself included, like the finely chopped version of the avocado base.  This version contains almost no sodium unless salt is added.

Rice is a common side dish in Tex-Mex cuisine.  At most restaurants, it is almost universally prepared in large batches ahead of time and cannot be modified to remove the generously added sodium.  But at home, you can make a great tasting, savory Tex-Mex rice dish with almost no sodium.  Here is my favorite Tex-Mex rice dish:

2 cups white or brown rice (I use brown jasmine rice)
3 cups water for white rice; 4 cups water for brown rice

2 cups diced onion - 13 mgs.
2 cups diced bell pepper – any color(s) -- 10 mgs.
1 cup chopped cilantro
2 Tablespoons Olive oil
Optional -- Jalapeno or other hot pepper to taste
Mrs. Dash™ Fiesta Lime seasoning, cumin, adobo, garlic powder, black pepper, ancho chili powder, smoky paprika, chipotle powder and other favorite southwest seasonings

In a large 5 quart Dutch oven, brown all of the vegetables in the oil.

Add rice and mix into the vegetables; cook for 2 to 3 minutes to coat the grains of rice in the oil.

Add water and bring to a boil.

Cover, reduce heat to low and simmer – for white rice, about 20 to 25 minutes; for brown rice, about 30 to 40 minutes – check to see if all of the water has been absorbed by the rice and that the rice is tender.

When there is no more liquid in the pot, and rice is at desired texture, remove from heat and fluff with serving utensil.

Makes 8 servings with a negligible amount of sodium per serving.

**Nutrition Information (brown rice):**
Serving Size: 1/8th of the total pot
Sodium:  3.6 mgs.
Calories: 231
Total Fat: 4.5 g
Saturated Fat: 0.3 g
Cholesterol:  0 mgs.
Carbohydrates:  42.1 g
Fiber:  3.5 g
Sugar:  3 g
Protein: 4.8 g

If this side dish is going to be served with numerous other dishes, the super low sodium level in the rice will help keep the overall meal within the planned amount of sodium. If there is available sodium for the meal that would permit the addition of salt to the rice, then add ½ teaspoon to the pot to increase the sodium to about 150 mgs. per serving, or ¼ teaspoon if you only have room in the sodium bank for a rice dish with about 75 mgs. per serving. No matter which level you choose, the rice will not contain the often huge amount of sodium found in grocery store mixes or res-taurant offerings, yet will have all of the flavors that we love from south of the border.

These same seasonings can be added to thin strips of chicken or beef to make the protein component for corn tortilla fajitas, or added to ground beef or ground turkey to make the protein filling for low sodium corn tortilla tacos. Build up the corn tortilla fajitas or tacos using small amounts of some (or all!) of the following sodium friendly ingredients:

Salsa - up to 50 mgs. per tablespoon
Ranchero Beans or Black beans -- 10 to 20 mgs. per tablespoon
Light Sour Cream -- 13 mgs. per tablespoon
Tex-mex Rice – nil
Chopped tomatoes -- nil
Shredded lettuce -- nil
Guacamole – nil
Grilled onions – nil

Grilled bell peppers – nil
Grilled mushrooms – nil
Sliced hot peppers – nil

Each nicely stuffed corn tortilla or taco could be assembled for about 100 mgs. of sodium apiece, a vast reduction from what is traditionally available elsewhere. Two or 3 pieces would certainly make a very filling meal.

These few recipe ideas represent just the surface of the Tex-Mex world. But many other recipes can be modified in the same fashion. Not everything will turn out to be a winner, but through brave experimentation you will be able to produce enough good dishes to satisfy your craving for this cuisine, and to even share it with friends and family. I often hear "Mmmm...this is very good. I am just going to add a little bit of salt to it." Personally, I am fine with that!

# XII.
# The Flavors of Asia

Oh, the dread of joining friends or family at a Chinese, Vietnamese, Japanese, Thai, Indian or other Asian restaurant. Some of these cuisines are among the saltiest on earth! The brown sauce used in a typical Asian stir fry may have several thousand milligrams of sodium per serving just from the soy sauce alone. And that's just the beginning! Many of these cuisines are generally thought of as healthy, especially because of their reliance on fresh vegetables, lean meats and fish. But between the excessive sodium and the fat content of the oil used for stir frying or deep frying (such as egg rolls), it is easy to take otherwise healthy ingredients and change them into a high fat, and even higher salt, nightmare.

But I love Asian foods. This certainly seemed like an area where my new low sodium regime was going to leave me high and dry. But over time, I began to experiment with homemade Asian food that looked and smelled pretty close to the real thing, and even tasted fairly good. This is one cuisine where it may be hard to fool your guests, as the lack of the bite of salt will be most noticeable. But the flavors of Asia are so varied and complex, that you can at least highlight the seasonings and flavors and turn out a much healthier version of what might otherwise be available in either the grocery store or a restaurant.

The sauce is the key to making an Asian inspired low sodium meal. Since traditional soy sauce contains up to 1,000 mgs. of sodium per tablespoon, and "light" or "reduced sodium" soy sauce still contains up to one-half of that amount, I have to take a completely different approach to developing an Asian-inspired sauce that will work for me. The great find has been several low sodium sauces in a variety of grocery stores that can be combined with other ingredients, and with each other, to produce a delicious Asian sauce. Here is a list of the low sodium sauces and glazes that I have found generally available in the stores in the greater Houston area, with the milligrams of sodium listed:

Whole Foods™ 365 Hoisin Sauce – 240 mgs. per tablespoon
Whole Foods ™ 365 Peanut Sauce – 200 mgs. per tablespoon

Marinade Bay™ Honey Ginger & Soy Sauce – 190 mgs. per tablespoon
Annie Chun's™ Shitake Soy Ginger Glaze – 180 mgs. per tablespoon
Ty Ling™ Orange Citrus Sauce – 115 mgs. per tablespoon
Panda Express™ Orange Sauce – 115 mgs. per tablespoon
House of Tsang™ Sweet and Sour Stir Fry Sauce -- 75 mgs. per tablespoon (Yes – 75 mgs.!)

These products have opened up a world of possibilities for Asian style cooking. I start with the basic recipe below for making any type of Asian sauce, and then modify the base by adding one or more of the products listed above in order to make various types of dishes.

### Pan Asian Stir Fry Sauce Base
To make four servings of an Asian dish, I use the following ingredients and quantities in a wok or large sauté pan:

4 teaspoons Toasted Sesame Oil
One-half teaspoon Liquid Smoke™
½ cup Pacific™ brand low sodium chicken broth -- 35 mgs.
Minced garlic, minced ginger, onion powder, mustard powder – all, or any combination, to taste. For more heat, add red pepper flakes.
Combine these ingredients over medium heat.

At this point, the mixture is ready to take on a flavor profile from one of the bottled sauces. I will use more quantity of the lowest sodium choices, and a lesser amount of the higher sodium choices. One of the best tasting sauces is made with the lowest sodium choice –

the sweet and sour stir fry sauce. I add 4 tablespoons to the base mixture.

One fourth of this now complete stir fry sauce contains only 84 mgs. of sodium, and is enough to make a full meal serving. You are now ready to stir fry. The entire dish can easily be prepared for under 200 mgs. of sodium per serving. The combination of toasted sesame oil and liquid smoke perks up the flavor of this sauce and helps to give it an Asian flair. To this sauce, any combination of protein and vegetables can be added. Most proteins add only 50 to 120 mgs. of sodium per serving, and the vegetables add a small amount of additional sodium. Thin strips of fresh, non- marinated chicken breast, beef, pork or duck are typical proteins. So are shrimp, scallops and chunks of various types of fish. Eggs, black beans and peas can all add to the protein. Each of these items is now readily available in a low sodium version. Eggs contain 65 mgs. of sodium each plus 6 grams of protein, so as a source of protein, eggs are not nearly as "sodium efficient" as poultry, beef, pork or fish.

A word of caution about frozen peas. I have purchased them in bags with zero sodium, and have also seen them in equally innocent looking bags that contained up to 300 mgs. per serving. This is another one of those warnings on careful label reading. Peas are more likely than almost any other frozen vegetable or legume to contain a high amount of added salt.

The vegetables to be used in a stir fry can be as creative as the cook's imagination will allow. Some of the very traditional items are onions, bell peppers, bean sprouts, bok choy, Napa cabbage, broccoli, asparagus, sliced carrots, celery, scallions, water chestnuts, bamboo shoots, baby corn, and many varieties of mushrooms. Some of these ingredients are often purchased in cans, such as the water chestnuts, bamboo shoots, baby corn and Asian mushrooms. Just as with the frozen peas, these items are more likely to be available in both low sodium and higher sodium canned options, and reading the labels is once again the key to making good choices.

To finish the dish, most people like to use either rice or noodles. Store bought Asian noodle preparations are often pre-seasoned and loaded with salt. The best choice is to make your own from plain boxed noodles, which entails boiling them ahead of time and adding them to the prepared dish. This will add virtually no sodium. Steamed white rice or cooked brown rice will also be a zero sodium addition if you prefer rice. Cook them separately without salt.

Given the above general guidelines, there is no reason why you cannot assemble a good tasting Asian inspired meal that contains anywhere from 200 to 350 mgs. of sodium per serving. This is incredibly low compared to traditional choices. A few of the Asian dishes in America's top restaurant chains have been highlighted for containing 4,000, 5,000 even 7,500 mgs. of sodium per entree! This is shameful – no dish needs to be shrouded and hidden in so much sodium. Let the other flavors speak for themselves without such a gigantic crutch. When restaurants are required to disclose to consumers more information about their dishes in an easily accessible format, I strongly suspect that these sodium levels will come down. I devoutly believe in personal choice, but an unsuspecting public cannot make choices if they don't even know just how ridiculously high some dishes are in sodium, fat, calories, sugar, and so on.

With a small collection of bottles in the door of your refrigerator, you can vary the Asian sauce to change the flavor palate of the meal. This basic sauce becomes a good Asian barbeque sauce by substituting a low sodium barbeque sauce discussed in Chapter X for the sweet and sour sauce. This actually reduces the overall sodium level of the sauce to 64 mgs. per serving. The peanut sauce can make a great Thai inspired meal, with crushed peanuts, lime and cilantro featured in the dish. The dark, rich and sweet hoisin sauce is more traditionally Chinese. The orange sauce is great with beef and shrimp dishes. Try all of the sauce and glaze offerings that you can find in your supermarkets where the per-tablespoon amount of sodium is under 300 mgs. and reduce the per serving amount by adding them to the Pan Asian stir fry sauce base I just described.

Armed with an array of homemade sauces, you can also look forward to joining friends and family at your favorite Asian eateries. I mix up a small container of sauce and bring it with me to the restaurant. I then ask for my meal to be steamed without any soy sauce, salt or added sodium, and no pickled vegetables. I mix my own sauce into the dish and I am very happy with the outcome. Of course, this still means no soup or other yummy starters from the appetizer menu, but it's a small price to pay in order to be able to walk into such an eating establishment and not ruin my low sodium goal for the day (or week!). Of course, there can be no sharing of plates with others; so if this is your habit at an Asian restaurant, you'll have to make that change as well. My wife Marjorie no longer cringes when I take out the small container of homemade sauce. The restaurant staffs at our favorite places never say a word about this, and I quietly take care of my own business while we all enjoy the evening together.

Indian food is typically centered around curry based spices and seasonings. You can do a lot of wonderful things with curry, cumin, cardamom, garam masala and a variety of peppers that don't require any salt, or need a very modest amount for the pot. Choose a fragrant basmati rice as a base for an Indian dish, which is available in both white and brown options. In addition to plenty of fresh vegetables, use lentils, peas and chickpeas for added protein, fiber, and enhancing the general bulk of the meal. Many people prefer to include meat in addition to the vegetarian ingredients. With generous seasoning, a great Indian dish can be made with a very low sodium content. There would be enough room left in the sodium bank to even add a small to moderate amount of salt to a pot for background flavoring. Try modifying any Indian or Central Asian recipe by removing most of the salt and adhering to the ideas in this paragraph. Here are two examples of very low sodium Indian dishes that are modifications to recipes that are popular throughout the sub-continent. The first is vegetarian, and the second recipe incorporates meat.

# Curried Peas

## Ingredients:
1 cup green peas
2 cups diced onion
1 cup diced  tomato (a canned product is fine)
1 green chili pepper (slit lengthwise)
¼ cup grated unsweetened coconut
4 Tablespoons vegetable oil
Spices and seasonings:
2 teaspoons fennel seeds
1 teaspoon turmeric powder
2 teaspoons red chili powder
1 teaspoon garam masala
2 teaspoons coriander powder
Curry leaves – a few, to taste (optional)

## Directions:
Cook the peas in boiling water until tender.

In a separate pot, combine grated coconut with 1 teaspoon of the fennel seeds, and grind into a paste.

Crush the remaining 1 teaspoon of fennel seeds and place into a skillet with the oil

Cook until they turn slightly brown.

Add the onions, green chili and curry leaves - saute until the onions are translucent

Add the tomatoes to the skillet and cook until they break down.

Add the turmeric powder, red chili powder, coriander powder and garam masala to the skillet, and cook for several minutes.

Add in the coconut and fennel seed paste and cook for a couple more minutes.

Add 2 cups water and bring the mixture to a boil.

After the water has cooked down by about half, add the cooked peas and saute until most of the liquid has been absorbed or cooked out.

Serve immediately.

**Nutrition Information:**
Serving Size: ¼ of pot
Sodium: 8.3 mgs.
Calories: 213
Total Fat: 15.9 g
Saturated Fat: 2.5 g
Cholesterol: 0 mgs.
Carbohydrates: 14 g
Fiber: 3.9 g
Sugar: 6.3 g
Protein: 3.4 g

If you prefer, add a ¼ teaspoon of salt to the recipe, and the per serving amount of sodium would increase to 156 mgs. per serving. Considering that this is a side dish, you may want to further adjust the sodium so that the overall meal you are planning fits within your guidelines. Lentils would increase both the protein and fiber content of this meal, either as an addition, or as a substitute to the peas altogether. If you don't want to bother with all of the different herbs and spices, a good substitute would be to just use curry powder and garam masala. Curry powder is available in a range of heat levels to suite your palate.

Main course Indian meals with sufficient protein can also be made with little or no added sodium. A basic masala recipe is an excellent starting point.  This recipe is very low in sodium, but bursts with flavor.

# Chicken Masala

**Ingredients:**
2 lbs chicken skinless boneless breasts or tenders
2 cups sliced onions
2 cups chopped cilantro
2 Tablespoons plain Greek yogurt
2 Tablespoons vegetable oil
2 teaspoons lime juice
Spices and seasonings:
1 fresh green chili pepper, diced (add more for more heat - one is mild)
2 Tablespoons cumin seeds
1 Tablespoon ginger
1 Tablespoon garlic
½ teaspoon salt
1 Tablespoon garam masala
2 teaspoons ground dried turmeric

**Directions:**
Make a marinade for the chicken by pureeing in a blender the chili pepper(s), cumin seed, salt, turmeric, ginger, garlic, cilantro and lime juice.

Add the yogurt into the blender and blend until very smooth.

Place chicken in a re-sealable plastic storage bag; add the marinade into the bag.

Make sure the marinade is evenly distributed through the chicken pieces.

Refrigerate for 2 hours or more (even overnight).

Heat the oil in a large frying pan over medium heat.
Stir in onions, and cook until translucent.
Add the chicken and marinade; bring to a simmer.

Reduce heat and cover pan – let simmer for about 30 minutes.

**Nutrition Information:**
Serving Size: 1/8th of pot
Sodium:  203 mgs.
Calories: 160
Total Fat: 4.5 g
Saturated Fat: 0.85 g
Cholesterol:  64 mgs.
Carbohydrates:  4 g
Fiber:  0.8 g
Sugar:  1.9 g
Protein: 26.9 g

I like to add fresh green beans to the pot during the final 10 minutes of the simmering process.  I cut them into thirds, and add about 1½ cups.  They are a great side accompaniment to this dish, and they absorb the flavors of the chicken and seasonings.  Green beans also add some fiber but very few calories and virtually no sodium.  So in my view, they just make the dish that much healthier.

Brown basmati rice is a natural accompaniment for this masala dish.  Use a 2 to 1 ratio of water to rice.  I add a few simple ingredients to the pot: a few drops of sesame oil, a pinch of cumin and a good sprinkling of Chef Paul Prudhomme's™ Magic Salt-free Seasoning.  Cover and simmer for about 20 minutes.  This adds no salt to the total, but of course, it does add calories, fiber, etc. So, with green beans added to the recipe served over rice, this is a complete low sodium Indian dinner.  And it is good!

# XIII.
# The Flavors of Italy

The flavors of Italy are among my favorite. Italian and Mediterranean cuisines are so full of flavor, but salt plays an important role in most of the dishes that we commonly know from this region. Removing most of the salt, but retaining the essence of these dishes, can be quite a challenge! This chapter represents another example of what it takes to modify and adjust traditional dishes while not losing the very best aspects of these great meals.

Most of the recipes in this chapter could be considered Italian comfort food. Many will rekindle memories of family dinners or weeknight casseroles in cool weather. Growing up, most of us paid no heed to the nutritional quality of the food we ate. We just knew whether we liked it or not. Recreating that memory, but with less than half the sodium, and much lower levels of fat, calories and sugar, remains a daunting challenge. Hopefully, the reader will be encouraged to take that challenge head on. These recipes might be a good starting point.

## Chicken & Pasta Marinara

Like many of the recipes in this chapter, this dish is hearty, warm and earthy. This recipe highlights the great flavor of Amy's™ Low Sodium Marinara sauce and the added protein and fiber in Barilla™ Plus High Protein pasta. This recipe makes 8 servings.

### Ingredients:
1 ¼ lbs skinless chicken breast -- 250 mgs.
8 oz. Barilla Plus™ High Protein elbows, rotini or any pasta combination -- 100 mgs.
1 can low sodium navy beans or other white/light color beans -- 50 mgs. per can
4 to 5 cups frozen mixed vegetables – corns, string beans, carrots, onions, peppers, celery, etc. -- 200 mgs.
1 jar (3 cups) Amy's™ Low Sodium Marinara sauce -- 600 mgs.
Optional – sliced mushrooms
2 Tablespoons olive oil
¼ teaspoon salt -- 590 mgs.
Seasonings to taste – garlic powder, onion powder, Italian herb blend, rosemary, oregano, black pepper

### Preparation:
Use a non-stick 5 quart Dutch oven to pan cook the chicken and a portion of the olive oil and seasonings.
Remove chicken to cool, then dice into small pieces. Leave remaining liquid in bottom of Dutch oven.
In a separate pot, cook pasta according to box instructions – about 12 minutes, then drain well.

Add frozen vegetables and rest of desired seasonings to Dutch oven.

Add all of the remaining ingredients to Dutch oven and reduce heat to simmer.

Let simmer for about 20 minutes, stirring occasionally.

Will make 8 main meal servings

Keeps in refrigerator for several days, and also freezes well.

**Nutrition Information:**
Serving Size: 1/8th of pot
Sodium:  224 mgs.
Calories: 318
Total Fat: 5.9 g
Saturated Fat: 0.6 g
Cholesterol:  40 mgs.
Carbohydrates:  40.2 g
Fiber:  7.9 g
Sugar:  5.4 g
Protein: 26.2 g

While I love to use Amy's™ Low Sodium Marinara in this and many other Italian dishes, a homemade version of the marinara sauce can further reduce the sodium content of the sauce by about one-half.  To make this version, I use one 28 ounce large can of tomato puree, one 14.5 ounce can of diced no salt added tomatoes, and a small amount (about ½ cup) of low sodium V-8™ Juice. Add diced onions and garlic, and add your favorite Italian seasonings.  Simmer on low heat until it thickens.  You can experiment with this sauce in many ways – add a splash of red wine, add thinly sliced mushrooms, shallots or leeks;

vary the level of heat with red pepper flakes, add thin sliced and chopped zucchini, and so on. I also like to add some of the low sodium chicken broth and then let it cook down to the desired thickness. It is good to have another option when you cannot find a store-bought product. Also, the homemade version will cost a lot less money to make. I use Amy's™ sauce very often because it is so convenient to just open a jar and there it is, ready to be used. However, it is quite expensive compared to most other pasta sauces. So when I have the time, or I am in the mood, I make my own. It takes about 30 minutes to make this homemade sauce. This is another item that can be made in a larger batch and placed into the freezer in containers. That would give you the great payoff of multiple servings for taking the time to make the sauce.

In lieu of chicken, this same marinara dish would be excellent with ground beef, chunks of stew meat or pork, ground turkey, and so on. There will be slight variations in the per-serving amount of sodium as you substitute different protein sources. I would vary the vegetables and seasonings and might even sprinkle on some shredded cheese in order to make a satisfying dish that still contains no more than 300 mgs. of sodium. If making a homemade marinara, the sodium content would be less and there would be enough room in the sodium bank for the added cheese. If you can serve your family or friends a tasty meal with pasta, sauce, cheese and a meat source, and still meet your sodium needs, what could be better than that?

Hot pasta dishes from the oven represent a classic style of Italian comfort food. Lasagna, manicotti, spaghetti pie, baked ziti, just to name a few. Below is one casserole dish with good overall nutritional quality that is also fairly easy to modify yet keep low in sodium (under 300 mgs. per serving). From this base recipe, other traditional casserole dishes can also be modified through experimentation.

## Tuna Noodle Casserole

This is a quintessential, everyday casual meal. It barely qualifies as Italian, but with pasta, cheese and Italian herbs, that's good enough to place the dish in this particular chapter. This recipe stays pretty close to the traditional elements of a good casserole, but manages to do so at under 300 mgs. of sodium. It is cheesy enough to satisfy, but loaded with protein and fairly skimpy on fat if you use tuna in water.

## Ingredients:

4 cans low sodium white tuna in water -- 280 mgs.

1 14 oz. box Barilla Plus™ high protein pasta – spiral, penne or elbows – or mix and match up to 14 ounces -- 175 mgs.

2 cups fat free high protein milk -- 250 mgs. (Note: regular fat free milk would reduce the protein content of each serving by 1 gram)

1 cup shredded pizza cheese or mozzarella -- 170 mgs. per quarter cup, for a total of 680 mgs.

2 cans Campbell's™ low sodium cream of mushroom soup -- 120 mgs.

1 ounce (5 ½ Tablespoons) shredded parmesan cheese -- 475 mgs.

I cup green peas (fresh or frozen) -- 20 mgs.

2 cups chopped onions -- 13 mgs.

2 cups chopped bell peppers -- 8 mgs.

Spices and seasonings:

½ teaspoon Tony Chachere's™ Creole seasoning -- 620 mgs.
Garlic powder, onion powder, black pepper, Italian herbs – oregano, rosemary, thyme, parsley, red pepper flakes

## Directions:

Place milk and mushroom soup into a large pot and bring to a simmer. Add in pizza or mozzarella cheese and cook until the cheese is fully melted and incorporated into the sauce. Then add all of the spices.

In a separate pot, prepare pasta in unsalted water according to the box instructions. It is preferable to slightly under-cook the pasta, as it will finish cooking in the oven.

Drain pasta and toss into a large mixing bowl.

Drain the cans of tuna and flake into the same mixing bowl.

Add peas, onions and bell pepper to the bowl.

Add sauce to the bowl and mix well to evenly distribute the sauce, tuna and vegetables.

Pour the mixture into a casserole dish sprayed with Pam™ or other cooking spray.

Sprinkle a thin layer of the parmesan cheese on the top of the casserole. You can also add a sprinkle of paprika if you prefer.

Bake in 350° oven for 30 minutes or until bubbly.

Remove from oven and let stand for 15 minutes.

Cut into nine equal size pieces.

Can keep in refrigerator for several days, and freezes well.

**Nutrition Information:**
Serving Size: 1/9th of casserole
Sodium: 293 mgs.
Calories: 364
Total Fat: 7.4 g
Saturated Fat: 2.4 g
Cholesterol: 40.9 mgs.
Carbohydrates: 44.2 g
Fiber: 6.8 g
Sugar: 7.8 g
Protein: 31 g

In addition to tuna, this recipe can be made using skinless boneless chicken breast or even chunks of lean beef. The sodium per serving will rise slightly for the beef, and fall slightly for the chicken, assuming you use products with the sodium content consistently discussed in this book. Turkey, lamb, pork, salmon and shrimp would also work well. If you change protein sources, consider varying the vegetables and spices as well, and build an arsenal of good low sodium casseroles. Sliced carrots go well with chicken and turkey. Thinly sliced zucchini is also a good choice. Add them to the onions and peppers, or mix it up and change everything. Try replacing the mushroom soup and milk sauce above with equal parts (4 tablespoons) all purpose flour and Smart Balance™ low sodium margarine. Then add 2 cups of the low sodium chicken broth to make a different sauce base for the casserole. Cook the flour in the margarine until it begins to brown, then add the chicken broth and stir over medium heat until a thick sauce begins to form. The shredded cheese will also melt into this sauce, but the overall flavor profile will be rather different than with the mushroom soup and milk-based sauce. I like both approaches, and I like to vary everything over time to keep these low sodium meals from getting too boring and predictable.

Here is one more hot casserole recipe that really satisfies your hunger for a traditional Italian meal like Mama used to make; well – as close to Mama's as possible under the circumstances. Thanks to all of the cheeses, this is one of the richer dishes in this book.

# Lasagna

## Ingredients:

10 pieces of Lasagna noodles (both the regular and no-boil noodles will work)

1 pound 95% lean ground beef -- 340 mgs.

1 cup diced onion

1 cup diced bell pepper

1 cup chopped parsley

2 Tablespoons olive oil

One half of a 15 oz. tub of Sargento™ part skim ricotta cheese

¼ cup egg beaters -- 115 mgs.

I jar (3 cups) Amy's™ Low Sodium Marinara -- 600 mgs.

Cheese Blend:
- ½ cup Pizza shredded cheese (170 mgs. per quarter cup, or 340 mgs.)
- 3 oz. low sodium mozzarella in water (20 mgs. per oz, or 60 mgs.)
- 4 oz. low sodium dry mozzarella (95 mgs. per oz., or 380 mgs.)
- 5 Tablespoons shredded Parmesan cheese (85 mgs. per tablespoon, or 425 mgs.)

Seasonings – onion and garlic powder, Italian herbs, Chef Prudhomme's™ Pizza and Pasta Magic, oregano, etc.

## Directions:

Heat olive oil in skillet.

Dice onions, bell peppers and ¾ cup of parsley; add to skillet.

Add seasonings to skillet, and then add in ground beef. Break up ground beef into very small pieces while cooking.

Boil lasagna noodles according to directions on box and then drain.

Blend ricotta cheese, egg beaters and remaining parsley in a separate bowl.

To create the cheese blend, shred the mozzarella cheeses and place into a mixing bowl with the pizza cheese and Parmesan. Blend all cheeses together.

To build the layers of lasagna:
- Start with one cup of the marinara sauce at the bottom of a 13 x 9 inch non stick baking pan.
- Add 3 noodles to bottom of the pan. Cut a fourth noodle in half and place on one end of pan to cover area not covered by the 3 noodles.
- Add one third of ricotta mixture.
- Add one third of ground beef mixture.
- Add one third of shredded cheese mixture.
- Repeat process two more times, using the other half noodle on the second layer on the opposite side of the pan.

Cover with foil and bake for one hour at 350°.

Cool before slicing into 9 equal servings.

Will keep in refrigerator for several days and also freezes well.

**Nutrition Information:**
Serving Size: 1/9th of casserole
Sodium: 286 mgs.
Calories: 360
Total Fat: 14.2 g
Saturated Fat: 7 g
Cholesterol: 63.6 mgs.
Carbohydrates: 29.9 g
Fiber: 2.8 g
Sugar: 6.9 g
Protein: 25.7 g

The sodium level per serving could be further reduced to about 250 mgs. by using the homemade marinara sauce described earlier in this chapter, in lieu of using Amy's™ Low Sodium Marinara.  You may not be able to find all of the specific cheese items mentioned in this recipe, but you should be able to find reasonable substitutes and develop your own cheese blend within these general sodium guide-lines.

Cold pasta dishes are popular and versatile for lunch and dinner, picnics and summer barbeques.  The following is a good representation of a flavorful, filling pasta salad with enough crunch and color to please your family and guests.

### Meat, Veggie & Pasta Salad

This recipe makes 4 good sized portions, with about 160 to 200 mgs. sodium per serving depending upon the selection of a protein source.

### Ingredients:
8 oz Barilla Plus™ High Protein pasta - elbows, rotini, penne, etc. -- 100 mgs.
8 oz grilled sirloin (200 mgs.), chicken (100 mgs.), turkey (150 mgs.), no salt added tuna (140 mgs.), etc.
4 Tablespoons Hellman's™ Canola Mayo -- 380 mgs.
1 cup cherry tomatoes – halved
1 cup diced red and green bell pepper
½ cup sliced scallions

½ cup diced red onion
1 cup diced celery -- 80 mgs.
1 cup diced carrots -- 90 mgs.
2/3 cup chopped parsley
2/3 cup chopped cilantro

**Herbs and spices to taste:**
- Garlic and onion powder
- Cumin, adobo, black pepper, smoky paprika
- Ancho chili powder, chipotle chili powder
- Curry powder
- Rosemary, oregano, thyme

**Directions:**
Cook pasta according to the box directions. Drain and cool.

Microwave the diced celery, carrots, onions and bell peppers for about 1½ minutes (just slightly tender).

Combine all ingredients in large mixing bowl - mix well to spread the mayonnaise throughout.

Chill and serve cold.

**Nutrition Information (made with chicken):**
Serving Size: ¼ of recipe
Sodium: 191 mgs.
Calories: 362
Total Fat: 7 g
Saturated Fat: 0.3 g
Cholesterol: 32 mgs.
Carbohydrates: 50.7 g
Fiber: 8.4 g
Sugar: 7.7 g
Protein: 24.9 g

I have varied this recipe in several ways. I sometimes like to add a can of no salt added black beans or garbanzo beans, and in doing so, I might even reduce the amount of animal based protein included in the recipe. Even if you don't adjust the meat protein, this addition can improve the overall nutritional quality of the meal – more fiber and protein with virtually no added sodium or fat, and if you were to reduce some of the animal protein, the benefits would be even greater. This is an especially helpful approach for anyone who is trying to manage cholesterol, but it also helps if you need to "find" some sodium reduction for the day. The fiber in beans is a top quality ingredient to help reduce cholesterol. In addition, a can of beans might have 30 mgs. of sodium in total, plus 20 to 25 grams of protein, and a similar amount of fiber. So replacing half the beef with beans, for example, would save 70 mgs. of sodium in this dish. That's not a lot, but every little bit helps. I also like the additional color and texture that beans add to a cold pasta salad. The other variations include adding different vegetables, like grilled zucchini or eggplant. Also, choosing a combination of spices and herbs that moves the flavor of the overall dish from Italian to Southwest, to Indian, Asian, and so forth is an easy step. Heat and smoke levels can also be adjusted, using grilled meats, red pepper flakes, and any other combination that you like.

# Pizza

The one sure thing that I thought was gone from my diet forever was pizza. I don't eat pizza all that often, but when you want it, you want it, and it's really nice to know that you can have it every now and then. That is what I thought I lost when I was told to live with 1,500 mgs. of sodium per day. Everything about regular pizza is salty – the crust, the sauce, the cheese, the toppings – everything! But hot, gooey, spicy, cheesy pizza is really so satisfying, that the thought of **NEVER** having it again just seemed like such a loss.

While learning how to make salt free bread, I came across a number of recipes for salt free pizza dough. It turns out, after some experimentation, of course, that they are pretty good, and can replicate the chewy texture and good taste of a traditional pizza dough as long as you don't crave to taste that salt bite that comes with the good old version of pizza. I have added herbs directly into the crust to help provide something savory to replace the traditional salty background flavor of regular pizza crust.

Once a decent crust was made, it didn't take long before some additional experimentation delivered a pretty solid pizza pie – a hearty meal in itself, served with a fresh salad.

# Salt Free Pizza Dough

## Ingredients :
½ package (1 ¼ teaspoons) dry yeast
1 cup of warm water
1 Tablespoon of olive oil (for the dough, plus some additional olive oil to brush on top of dough prior to placing in the oven)
1 Tablespoon of sugar (you could use less and see if you are satisfied with the result)
2 cups of white all purpose flour
Optional:   herbs and seasonings worked into the pizza dough (rosemary, salt-free Italian seasoning, oregano, garlic powder, etc.

## Directions:
Pour ¼ cup of the warm water in a small bowl. Add the sugar and yeast. Let stand for 10 minutes.

Pour the flour in a medium size bowl.

After 10 minutes, add the rest of the water to the yeast-sugar mixture and mix.  Add this to the flour. Mix lightly while adding the oil.

Knead lightly on a flat, floured surface for a couple of minutes. You can always add a bit of flour or water if needed.

If using herbs and seasoning, add into the dough at this time.

Put the dough in the bowl, cover and let stand in a warm place for 45 minutes.

After 45 minutes, turn out dough onto a flat, floured surface and using a rolling pin, flatten until the dough fits your pizza pan. Use more flour if needed to avoid sticking.

Press the dough in the pan, and make an edge.

Brush the top of the dough with the reserved olive oil.

Bake the dough without any of other ingredients for about 8-10 minutes at 400°.

Remove crust from the oven and you are ready to add the toppings to make pizza.

**Nutrition Information – Pizza Dough only:**
Serving Size: 1/8th of pie
Sodium:  1 mg
Calories: 139
Total Fat: 1.8 g
Saturated Fat: .1 g
Cholesterol:  0 mgs
Carbohydrates:  25.9 g
Fiber:  0.9 g
Sugar:  1.6 g
Protein: 3.6 g

Now let's make pizza using this crust.

# Low Sodium Pizza

**Ingredients:**
Salt-free pizza dough, prepared as described above -- 1 mg.
1 cup Amy's™ Low Sodium Marinara -- 200 mgs.
8 ounces of your choice – chicken (100 mgs.), turkey (150 mgs.), ground beef – 95% lean (170 mgs.) – chosen for this recipe
1 cup shredded pizza cheese (170 mgs. per quarter cup, or 680 mgs.)
Your choice: I always use diced onions and bell peppers, but also consider adding mushrooms, tomatoes, other veggies, garlic, etc.
Spices and herbs: oregano, no salt Italian seasoning blend, rosemary, garlic powder, Chef Prudhomme's™ salt-free Pizza and Pasta Magic, etc.

## Directions:

While pizza dough is rising, brown the ground beef and diced vegetables in the olive oil with Italian seasonings of your choice. Make sure that beef is broken into very small bits, for even distribution on top of the pie.

Take the pizza dough out of the oven.

Add marinara sauce to the dough, then add the meat and other toppings, and finish with the cheese. Spread as evenly as possible, and make sure that a good portion of the total ingredients does not end up toward the very center of the pie – this will overweight the end of a slice and make it more difficult to handle.

Place in oven for about 15 to 20 minutes at 400° or until browning begins to occur on the cheese and crust.

Remove from oven and let stand for about 10 minutes.

Cut the pie into 8 slices.

## Nutrition Information (Includes the pizza crust and choice of ground beef):

Serving Size: two slices (¼ of pie)
Sodium:  267 mgs
Calories: 567
Total Fat: 19 g
Saturated Fat: 5.8 g
Cholesterol:  54 mgs
Carbohydrates:  63 g
Fiber:  4.8 g
Sugar:  8.7 g
Protein: 28.5 g

This is a good basic recipe for individual experimentation.  There are many options.  Chef Paul Prudhomme's™ Pizza and Pasta Magic will add a cheesy and spicy kick without a drop of sodium. It's a great topping for this pizza. It even has the aroma of parmesan cheese! You may want to vary the amount of meat or cheese. Regular shredded pizza cheese can be found with about 170 mgs of sodium per quarter cup (many are higher, so be careful), but a mere 2 tablespoons of Parmesan would contain roughly the same amount of sodium as the quarter cup of pizza cheese. You might want to sprinkle a small amount of real Parmesan cheese on the pizza for added flavor, or consider some kind of mix between the different types of cheeses. If you are going to experiment, keep close track of the sodium bank for the entire pizza. If two slices will be sufficient for a meal, and you are aiming for about 300 mgs. from the pizza portion of the meal, then a small amount of Parmesan cheese added for the whole pizza will be fine. While I have not found a tastier commercial pasta sauce for the amount of sodium used than Amy's™ Low Sodium Marinara, you may want to try and make your own pizza sauce by using the marinara recipe described earlier in this chapter. By reducing the sodium with the homemade sauce, there could be a bit of added room for even more cheese. Of course, all of this added cheese will ramp up the calories, fat and cholesterol, so keep track of this if any of these things are important to you.

# XIV.
# Happy Thanksgiving

Who doesn't love Thanksgiving?  Gathering with family and friends.  The fantastic smells emanating from the kitchen.  Great football rivalries on the television.  The colors of fall, the crackle of the fireplace, the mounds and mounds of unhealthy and highly salted food.  Ahhh........ Wait – that last point makes Thanksgiving one of the *hardest challenges* for people who are on any kind of dietary restriction or limitation.  It is not a holiday about limitations.  It is about giving thanks for the bounty that we enjoy. Well, there's usually too much bounty on the table from a health standpoint.  No one wants to be a Thanksgiving party pooper, but this holiday requires some serious planning and adjustments if you intend to stay reasonably on course for reaching or maintaining your low sodium goals.

It may be lacking in some of the fancier flourishes of the season, but here is a basic Thanksgiving "starter menu" that can meet the goals of the traditional holiday meal while at the same time meeting your low sodium needs at around 300 mgs:

*Roasted Turkey-- 75 mgs.*

*Oven Baked Stuffing – 50 mgs.*

*Sweet Potatoes – 75 mgs.*

*Green Beans – 20 mgs.*

*Cranberry Sauce – 40 mgs.*

*Herbed Crescent Rolls and Buttery Spread – 45 mgs.*

## Roasted Turkey

The challenge is to purchase a turkey with minimal processing and no added sodium. In many parts of the country, a fresh turkey can be obtained. Turkey in its natural state is likely to contain between 50 to 75 mgs. of sodium per 4 ounce serving, which is right in line with most of the other main protein sources mentioned in this book. A turkey with this amount of sodium is available in some grocery stores. An example is the Diestel™ brand from California, available at Whole Foods™ and many smaller natural foods stores. The Diestel™ web site (*diestelturkey.com*) contains a store locator link that may be helpful. Other brands are available, but make sure to ask your grocer to provide you with the nutrition label if it is not already affixed to the packaging. A so-called "natural" turkey can still have added sodium, and I have seen labels on turkeys ranging from 50 mgs. per 4 ounce serving, up to 500 mgs. per 4 ounce serving! This is definitely another case of buyer beware and read the label. If you cannot find a turkey with a nutrition label in the low end of the sodium range, ask the butcher at any store for organic, minimally processed turkey with no added liquid solutions. In most cases, you can rely on what you are told by the butcher. If he or she gives you a clueless look, however, move on to the next store and try again.

For added flavor, I love to roast poultry with a simple insertion into the cavity – celery stalks with the leaves on, an onion cut in half or quarters, several cloves of garlic, and a few lemon halves. I also like to add a few springs of fresh rosemary. This combination provides a great flavor and aroma combination that seeps throughout the bird. For seasonings on the skin, use your favorites. I like sage, garlic powder, onion powder, black pepper and thyme. I also love to add paprika, but not everyone likes the flavor that it imparts. If you have a Thanksgiving secret spice blend, go for it! If it used to contain salt, just try it without the salt and see if your taste buds have evolved.

**Thanksgiving Stuffing**

I prefer to bake stuffing in a separate baking pan or just make it on the stove top, mostly because i do not want to have the stuffing absorb a gallon of turkey fat. Yes, that would taste very good, but my health goals extend well beyond sodium reduction. Also, this allows the cavity to be stuffed with the fragrant and savory items mentioned above, which do make for a more flavorful turkey. This stuffing is a basic recipe to which many additions could be made.

**Ingredients:**
One loaf of no sodium white bread, or one loaf of low sodium Ezekiel™ bread – mixing and matching is perfectly acceptable
2 cups chopped celery
2 cups chopped onions
3 cups Pacific brand low sodium chicken broth
5 Tablespoons Smart Balance low sodium spread
Spices and herbs – sage, black pepper, garlic powder, onion powder, savory, celery seed, or any other spices that suit your tastes

**Directions:**
Cut the bread loaf into sandwich slices, and then cube the slices.

Place cubes on a baking sheet and bake in oven for 20 minutes at 325°, or until brown and hard to the touch (this is to remove moisture from the bread). Parchment paper makes it easier to avoid sticking, but it is optional.

Place chicken broth and margarine spread into a large pot and bring to a boil.

Add bread and seasonings to the pot of broth and margarine and stir to fully incorporate into the liquid.

To finish on stove top – cook until moisture level is satisfactory to your tastes. It will start out very moist, and needs to absorb liquid and experience some evaporation.

To finish in the oven - turn out onto a casserole pan and spread evenly. Do not pack tightly. Bake in 350° oven for about 30 to 40 minutes. The top should be golden and have a light crust forming.

Remove from oven and cool for 10 minutes before serving.

**Nutrition Information**
Serving Size: 1/12th of casserole dish

Sodium: 40.4 mgs.
Calories: 142.7
Total Fat: 5.3 g
Saturated Fat: 1 g
Cholesterol: 0 mgs.
Carbohydrates: 19.4 g
Fiber: 1.1 g
Sugar: 2.9 g
Protein: 3.4 g

As for additions and variations to customize the flavor of this dish, I would add chopped walnuts for starters, and possibly small diced carrot pieces that have been pre-cooked. I really like sliced mushrooms in the stuffing. Some people add cranberries or other fruits to their stuffing, combining sweet and savory like they might with other holiday recipes. So far, the listed additions would add practically no additional sodium per serving. By contrast, the average prepared stuffing mix contains between 500 to 1,000 mgs of sodium per prepared serving, so this represents a huge sodium savings. If the low sodium recipe is too low for your tastes, and there is room in the sodium bank for the meal, a quarter teaspoon of salt added to this dish would take the per serving sodium level up to 100 mgs. This type of adjustment is always a good idea for experimentation. It may be that your taste buds need more salt in the stuffing, and less salt somewhere else, in order to avoid a taste bud revolt. Try adding or removing sodium from any of the recipes in this book and make your own adjustments, bearing in mind the sodium goal for your meal as well as your daily goal.

## Herbed Crescent Rolls

These rolls are a tribute to my wife, Marjorie, and her family's tradition of serving homemade warm, soft buttery crescent rolls at Thanksgiving. Of course, this recipe is a modification from the super buttery and much saltier original version. These soft, aromatic rolls start with the same basic bread recipe in Chapter III, and can be made from either the bread machine or the stand mixer on the dough hook. The changes in the recipe start at the point where the bread is ready for its final rise before baking. If using the bread machine, that moment is when the cycle reaches the point of placing the dough into a bread pan or to make rolls. If using the stand mixer, that point is reached after the second 45 minute rise.

## Directions:

While waiting for the dough to finish its second rise, melt 4 tablespoons of Smart Balance™ Low Sodium Buttery Spread in a small saucepan. This may seem like a lot of margarine, but this recipe will make 32 crescent rolls, so a very small amount will end up on each roll.

Turn out the dough on a lightly floured surface and cut into 2 equal halves, each one weighing approximately 12 to 12.5 ounces. I use the electronic kitchen scale to make sure that I am measuring accurately. Using a rolling pin, roll out the first half very thinly, so that it encompasses a circle with about an 11 or 12 inch diameter. The dough will be very thin at this point. I use a rolling mat with a guide printed right on it to show me the correct size. Tupperware™ makes one that can be purchased online. You can also just use a ruler, or if you are better than me, you can eyeball it.

Bring the melted spread to the area where the rolls are being prepared, and, using a pastry brush, fully coat the entire circle of dough with a thin layer of the melted spread. As an option, you can sprinkle on a modest amount of Mrs. Dash™ Garlic and Herb seasoning, or any favorite herb and seasoning combination or blend. Coat evenly. I have tried using a heavier sprinkle of these ingredients, and when I do, the dough does not seem to rise as well. So I stick to a fairly light coating.

Using a pizza slicer or a knife, cut the circle into 2 halves, and then cut each half into 2 more halves. Continue halving until there are 16 thin slices of dough encompassing the circle. They may look very thin, and they may not be completely uniform, but that is perfectly fine for this recipe. Notice that as you slice through the dough it separates quite easily into workable wedges. Starting with the widest part of a wedge of dough, begin rolling the dough wedge onto itself until all of the dough has been completely wrapped. Place the end tip face down on a parchment covered baking sheet. Each of the 2 dough halves will fill one baking sheet with 16 little crescents. Place them as far apart as possible on the sheet, as they will rise considerably during the final proofing, and will rise even more once they are placed in the oven to bake.

After the first baking sheet is filled, lightly brush each crescent roll on the sheet with a small amount of the melted margarine spread. Put the baking sheet uncovered into a draft free environment like the lower oven of a double oven, or cover lightly with cling wrap and keep in a warm part of the kitchen away from drafts. I turn the oven on for a few minutes and then turn it off before adding the rolls. The extra warmth helps with the rise.

Repeat the process for the second dough half, and the second baking sheet will be completed about 5 to 7 minutes behind the first one in terms of its final proofing. This is not a problem.

Preheat oven to 375°. When the first baking sheet is finished proofing (about 45 minutes), place it in the oven for about 12 minutes, or until the rolls start to turn a golden brown color. When the second baking sheet is ready, add it to the oven as well. Remember the time difference between the two, although using the golden color as a guide is probably the best indication of when they are done. Remove from baking sheet onto a cooling rack and let them cool completely. Once they are cool, transfer into a plastic bag and let them rest overnight before serving. This last step gives the rolls a chance to develop a softer, chewier texture, and the smell of the fresh baked bread with the herbs and garlic is just outstanding. When I have sampled a roll prior to this resting step, it was good, but after this step, they were much better.

Each of these herbed crescent rolls has less than 5 mgs. of sodium! Two or three on a Thanksgiving dinner plate makes for a perfect complement to the rest of the low sodium meal. If you have any additional sodium in your sodium bank, you could add ¼ or ½ teaspoon of salt to the dough mixture, and the rolls will probably taste even better. At ½ teaspoon, each roll will have about 78 mgs. of sodium, and at ¼ teaspoon, each roll will have around 40 mgs. Another option is to use a regularly salted margarine or butter, which, without any other added salt, would take each roll up to about 12 mgs. Of course, that is a lot less than adding salt to the dough. I recommend that you Experiment with this ahead of the holiday and determine which choice works best for you. Remember, you are not likely to be limiting yourself to just one roll, so the sodium can add up pretty quickly. I am happy with the super low sodium rolls, to which I add a bit of the low sodium Smart Balance when I eat them. Combined with the herbs, there is an excellent overall flavor! Smart Balance™ also makes a 90 mgs. of sodium per serving margarine, and it would work well in producing rolls with about 12 mgs. of sodium per roll.

## Sweet Potatoes

Good news. This is an easy one. A baked or mashed sweet potato has a bit of naturally occurring sodium (between 50 to 75 mgs. per serving), but most of the traditional additions are very low sodium or contain none at all. Consider the following:

½ cup miniature marshmallows – 20 mgs.

2 teaspoons brown sugar – 1.6 mgs.

1 Tablespoon maple syrup – 1.8 mgs.

Sodium is not the problem with these potential additions – sugar is the problem! You can really load up on sugar considering that the sweet potato is a starchy tuber to begin with and therefore, does contain a fairly large amount of sugar itself. However, most research suggests that the plain sweet potato may actually be beneficial to diabetics, so it is the souped up, fully loaded traditional holiday version of sweet potatoes that is of questionable benefit. Personally, I go easy on the added sweeteners, and use some Smart Balance™ low sodium spread, cinnamon and chopped pecans or walnuts to make my Thanksgiving sweet potatoes. Another good idea is to take the cranberry sauce that is already on your plate, and move some of it onto the sweet potatoes. This is a great flavor combination, and you've probably already had both of these items in your mouth at the same time anyway. The sweetness from the sauce is really perfect on a plain baked sweet potato, or even if they have been mashed with margarine and cinnamon, nuts, etc.

## Green Beans

This is one of the healthiest vegetables available, full of good nutrients and naturally very low in sodium – about 7 mgs. per cup. I ensure no sodium hanky panky by picking fresh beans from the produce section, although frozen beans would also work. I steam them in a covered container in the microwave with just a small amount of water. At this point, they are a blank green canvas ready to be painted. All of the herbs and spices mentioned throughout this book could be applied to the green beans, and I also like to add in slivered almonds. The Smart Balance™ low sodium spread is excellent on vegetables, imparting just a hint of saltiness. A teaspoon per serving would add just 10 mgs. So a half cup serving of green beans would contain

17 mgs. of sodium, and a full cup of beans with the same 1 teaspoon of margarine would contain 20 mgs. I like to finish the green beans in a large sauté pan, where I might add a splash of the low sodium chicken broth as well. I want to get every little bean nicely coated with all of the ingredients.

The result would be roughly the same for just about any green vegetable chosen, including Brussels sprouts, cabbage, zucchini, and just a little bit higher for broccoli, spinach and cauliflower.

## Cranberry Sauce

Whole berry or jellied, home made or store bought, the basic stuff is very low in sodium, figure 20 mgs per serving. Many people like to doctor their cranberry sauce with all sorts of things. I like whole berry sauce right out of the can with added pineapple tidbits and walnut pieces, neither of which would add any sodium to the dish.

What's missing from the meal? Well, gravy for one. I have yet to find a good tasting low sodium gravy worth adding to my plate. Mashed potatoes are missing, since the sweet potatoes are the low sodium alternative. Most people put the following yummy higher sodium things in mashed potatoes – milk, butter, salt – just to name a few, which takes an otherwise very low sodium tuber and cranks it up to potentially hundreds of milligrams per serving! As much as they are a beloved holiday tradition, there is a good alternative to these salty additions. If you can be happy with very low sodium mashed potatoes, then by all means, serve them. Some additions that may help are the low sodium margarine, black pepper, garlic, onion powder, sour cream, yogurt, or a small amount of butter-milk, and finely diced chives. As with any of the items on your Thanksgiving table, the other diners can add salt from the shaker if they feel the need to do so. Most will! But you have nonetheless mastered a low sodium meal which most people can't begin to even think about modifying on their own.

Yet one final item in absent from this Thanksgiving table – dessert! A chapter on desserts is coming right up, but for Thanksgiving, I offer the following guideline. If you've been able to keep to the menu detailed in this chapter, and if you've been reasonably good and consistent throughout the year, then this is one of those moments when the best advice for your long term sustainable low sodium plan is to have whatever your heart desires from the traditional, fully loaded dessert selections – apple or pecan pie, pumpkin or sweet potato pie, cookies, cake, ice cream, whipped cream topping, etc. – and don't worry about the sodium or other nutrient components in the dish. You've got to have a few exceptions to even the best, most sensible rules. This is one of those times. So Enjoy!

# XV.
# Just Desserts

There is good news in this chapter - there are many dessert choices with low to moderate amounts of sodium. Some are pretty obvious, like fresh fruit, for example. Others might surprise you. But a sweet ending to a meal doesn't have to be a salty experience. Fruit may be the best example – any fruit can be sliced, diced and pureed into a nearly sodium free dessert. My favorites include berries of all types, pineapple, watermelon, grapefruit and grapes.

Nuts are another way to enjoy a good ending to a meal without added salt. Practically every variety of nuts is now available raw, or roasted without salt. I like them all. Nuts can also be combined with fruits to make an interesting compote. Watch out for the high fat content of nuts, but adding a few nuts to a fruit dish is a really good idea. As an added benefit, both nuts and many fruits contain solid amounts of fiber as well.

Perhaps the next grouping of desserts may be a bit of a surprise. Virtually every frozen treat category has a large number of low to moderate sodium choices. For example, whether you prefer super premium high fat ice cream, regular ice cream, low fat or low/no sugar added ice cream, frozen yogurt, frozen custard, gelato, sherbet, sorbet, or water ice – a cold treat is available with a very reasonable amount of sodium. For dairy-based desserts, the typical range of available lower sodium options is between 40 to 80 mgs. per half cup serving. If you have typically selected something well above this range in the past, you may want to see if there is an alternative that you like but that contains less sodium. Without too much difficulty, you should be able to find good substitutions.

Some people like to add a whipped topping to their dairy, fruit or icy dessert. Cool Whip™ adds only a paltry 5 mgs. of sodium per 2 tablespoon serving. Redi Whip™ and similar whipped cream products range from 0 to 10 mgs. of sodium per a 2 tablespoon serving. Other possible additions could include fruit preserves, most of which have zero or negligible sodium, or Nutella™, a rich chocolate hazelnut spread with only 15 mgs. of sodium per 2 tablespoon serving. Many chocolate syrups also contain the same low 15 mgs. of sodium per 2 tablespoon serving. Of course, the sugar and fat can accumulate quickly if you combine the more decadent ingredients from this list.

Cheese is another popular item to place on a dessert plate. Slices of the low sodium mozzarella in water can be served with fruit, or can be topped with jellies or preserves and served on a cracker. There are low sodium cracker choices mentioned elsewhere in this book, such as Hint of Salt Ritz™ Crackers. The Cheesy Girl™ goat cheese and similar products could be substituted for those who prefer a tangy cheese taste combined with fruit.

Puddings are generally higher in sodium, starting around 130 mgs. per serving and rising from there. This is true of all flavors of stovetop cooked or instant pudding, rice pudding, tapioca pudding and similar products. Local options may exist for lower sodium varieties, but I have found very few. One national brand, Jell-O™, does make a sugar free chocolate mousse that contains only 100 mgs. per serving cup. It is reasonably tasty by itself, or could be used as an ingredient in other dishes. Other flavors of this mousse product have higher levels of sodium. Gelatin desserts, such as Jell-O™ brand gelatin, typically start at around 45 mgs. per serving, and depending upon the brand and flavor, rise to 150 mgs. or higher. But many flavors can be found in the 45 to 65 mg. range. So, a bowl of "jello" and whipped topping can be assembled for as little as 50 mgs. of sodium. Kids love this type of dessert, with squiggly jello and creamy topping. They don't have to know that they are making a good choice in terms of sodium. If you select the sugar free jello, this dessert can be essentially fat free, extremely low sugar, and low sodium, and yet it has a mainstream taste that doesn't feel like a deprivation.

It gets more complicated when moving to cakes, pies and cookies. Most recipes call for salt, baking soda or baking powder. These ingredients typically knock these desserts out of the acceptable low sodium list, unless they are capable of being reasonably modified. However, you can in some instances control the level of sodium. There is a good salt-free baking powder call Featherweight, made by Hain™. Substituting this product for the real thing tends to do little to change the overall flavor and texture of baked goods. There is also a no salt added baking soda which can be substituted for the traditional thing -- Ener-G™ brand baking soda. This product requires using twice the amount of regular baking soda called for in the recipe. Both of these items can save a lot of sodium. Both of these substitute products are available at some health food stores as well as online at sites such as *amazon.com*.

I do not recommend using salt substitutes in lieu of salt, however, as they lack the right chemicals for the needed reaction. To make matters worse, they may impart an aftertaste to most dessert items that most people would find unacceptable. Instead, focus on recipes that call for reduced amounts of salt or eliminate it altogether.

There are plenty of low sodium baked dessert recipes worth trying. *A note of caution* – many low sodium desserts are made with a hefty amount of sugar, and quite a few also have a high fat content. Making a lower sugar, lower fat, low sodium dessert that also tastes good is a particular challenge, but one worth taking on in the quest for a low sodium lifestyle. Here is one good example:

# Peach, Apple (or other fruit) Crumble

**Ingredients:**
2 cups peeled and sliced fresh fruit
½ cup graham cracker crumbs (varies by brand, but graham crackers can be found with about 120 mgs. per whole cracker)
1 Tablespoon melted Smart Balance™ Low Sodium spread -- 35 mgs.
½ teaspoon cinnamon
⅛ teaspoon nutmeg
Pam or other cooking spray

**Directions:**
Preheat oven to 350°.

Spray Pam™ on an 8 inch square baking dish.

Layer the fruit on the bottom of the dish.

In a separate bowl, combine melted margarine, graham cracker crumbs and spices into the crumble mixture. Spread crumble over the fruit.

Bake for about 30 minutes.

**Nutrition Information (using fresh peaches):**
Serving Size: ¼ of recipe
Sodium:  100 mgs.
Calories: 92
Total Fat: 2.8 g
Saturated Fat: 0.5 g
Cholesterol:  0 mgs.
Carbohydrates:  16 g
Fiber:  1.5 g
Sugar:  8.5 g
Protein: 0.7 g

This makes a warm, tasty dessert with a reasonable amount of sodium, sugar and fat per serving.  This could be topped with vanilla ice cream or whipped topping (or both!) and still be an acceptable choice for a low sodium dessert, although this would necessarily make for a sweeter and richer dessert.

The key principle in the crumble dessert is to limit the bread or grain based ingredients which are likely to contain the most amount of salt.  Using the same principle, almost any pie made with a thin graham cracker crumb crust would be a good choice.  Layers of sliced fruit or berries, whipped topping and chocolate hazelnut spread could be combined into a very nice dessert pie. Bake the crust and fruit together, then cool completely before adding the other toppings. Try making the same crumble recipe with dry rolled oats (the 5 minute cooking type) and the sodium content will be reduced to almost nothing. To make this palatable, I would add 2 to 3 tablespoons of brown sugar to the recipe since you will also lose the sweet taste of the graham crackers in addition to the sodium. Oatmeal crumbles are tasty in their own right, and what a bargain when it comes to sodium!

Of course, I will save up some sodium milligrams and splurge on a dessert when the right occasion calls for it. I previously discussed this in the Thanksgiving chapter. Special events like birthdays and anniversaries, New Year's Eve, and the like, could all involve a higher sodium dessert to cap off a good celebratory meal. A great restaurant dessert may also be worth saving up the milligrams. At least with dessert, you are far less likely to go way overboard on sodium content compared to main entrees, appetizers and soups.

# XVI.
# Sunday Brunch

Do you like to put together a variety of dishes for a lazy Sunday late morning? Are friends or family joining you for a casual celebration? This chapter will present an assemblage of a variety of low sodium items that could be part of your inventory of dishes to prepare for any day of the week, but would certainly be a welcome sight at a Sunday brunch. I have a bit of a psychological accounting trick that I use for brunch that helps me really enjoy the meal. If I am going to skip breakfast and have a good sized brunch in the late morning, then I am fine with adding together the sodium bank amounts from both breakfast and lunch, provided I am going to stick to the meal plan for the rest of the day, and have an afternoon snack, plus dinner. If I can commit to do this, then my sodium bank for brunch will be in the range of 800 mgs. That gives me a lot of room to enjoy some of my some favorite dishes, and to not let sodium get in the way of a nice meal with family or friends.  This really isn't much of a trick – it's more of an attitude adjustment for just that one day.  Thinking of the sodium allowance in this way seems to work for me.  Try it and see whether or not it is a helpful technique for you.

Here are some sample choices to consider when assembling a brunch menu:

Mejadrah (recipe below)

Hash Browns (recipe below)

Fritatta (recipe below)

Oatmeal (see below)

Black Olives (no salt added)

Eggs with Grits (see below)

Tuna Salad (from Chapter IV)

Corn Bread Muffins (see below)

Mediterranean Tortilla (corn tortilla, cheese, mejadrah)

Yogurt, Fruit and Nuts (from Chapter IX)

Fresh Fruit and Berries

Fruit Crumble (from Chapter XV)

## Mejadrah

What is this dish, you might ask?  Mejadrah takes several very ordinary ingredients and combines them into something seemingly much better than the sum of its parts. Likely originating in Egypt, mejadrah features the ancient grains and flavors of the Middle East. Exotic yet earthy, this is a great dish for a large gathering. My wife Marjorie has a knack for making this taste so good that some guests find it irresistible. This recipe, which halves the amount of salt that she typically uses, still tastes great. It makes 10 hefty side dish servings, and can be stored for 3 to 4 days in the refrigerator. It also freezes very well.

## Ingredients:
2 medium onions – cut into thin strips
1 cup uncooked green lentils
1 cup uncooked rice – jasmine or basmati white rice work best, but brown rice is also a good choice
4 Tablespoons olive oil
1 to 2 teaspoons ground cumin (to taste)
Dash of pepper
½ teaspoon salt (1,180 mgs.) – the only appreciable source of sodium in the recipe

## Directions:
Boil 5 to 6 cups of water for later use.

Boil lentils in 3 to 4 cups of water for about 20 minutes.

While lentils cook, heat 3 tablespoons olive oil in a large (5 to 6 quart) sauté pan. Add the onions and cook over medium heat until they have carmelized. Stir onions frequently while cooking. Brown edges are fine. This should take about 30 minutes.

Drain the lentils.

Add the remaining one tablespoon of olive oil to the large sauté pan and onions, and stir in the uncooked rice.  Once the rice is toasted (about 2 to 3 minutes), add the cooked lentils, cumin, salt and pepper.  Add about 2 cups of boiling water.  Stir it all up and cover, letting the rice absorb the water.  Let simmer for about 5 minutes over low heat, then check and add more water.  Repeat this process.  You should end up using about 4 to 5 cups of water overall.  The rice and lentils should be completely cooked but you don't want the lentils to fall apart.  This part of the cooking process takes between 20 to 30 minutes.

Spoon the mejadrah onto a large serving dish for immediate use, or place into storage containers.

Recipe makes about 10 side dish servings.  I like it best when served hot, but some people like it served warm, or even at room temperature (my wife).

**Nutrition Information:**
Serving Size: 1/10th of recipe
Sodium:  122 mgs.
Calories: 208
Total Fat: 5.8 g
Saturated Fat: 0.4 g
Cholesterol:  0 mgs.
Carbohydrates:  31.9 g
Fiber:  7.4 g
Sugar:  3.2 g
Protein: 6.8 g

When served with a choice of meat and a salad, this dish helps to make a complete meal packed with protein and fiber, but light on sodium and calories. It is a great side dish for a Sunday brunch – savory, with great texture and very flavorful. People can take as little or as much as they want. It's just an easy staple for a large gathering.

## Hash Browns

This is one of those potato products that can easily be purchased frozen in a variety of ways. The best chance for a near-zero sodium product is plain frozen shredded potatoes, or plain frozen diced potatoes. There are also numerous versions with almost no sodium where the diced potato is accompanied by diced onions and bell peppers. These are the best choices for a good pan of low sodium hash browns. The same caution is needed in this section of the grocer's freezer case as with the French fries or steak fries – every bag should be inspected to verify the actual nutritional information. There are hash brown products sold with hundreds of milligrams of sodium, often on the same shelf of the freezer case.

This is a simple recipe, and should be familiar to most people who cook potatoes. The amount of oil has been reduced, and the sodium has been all but eliminated. Otherwise, this is a straightforward potato recipe.

## Ingredients (makes 4 servings):
3 cups shredded or diced potatoes
½ cup chopped onion
½ cup chopped bell pepper
1 Tablespoon plus one teaspoon olive oil
Seasonings – smoked paprika, garlic powder black pepper, cumin, herbs such as rosemary, oregano, thyme, herb blends, etc., to taste

## Directions:
Brown onions and peppers in olive oil in a non stick pan or cast iron skillet

Add potatoes and seasonings – cook until the potatoes are a deep golden brown color, mixing the contents occasionally as they cook.

## Nutrition Information:
Serving Size: ¼ of recipe
Sodium:  13 mgs.
Calories: 99
Total Fat: 4.7 g
Saturated Fat: 0.3 g
Cholesterol:  0 mgs.
Carbohydrates:  13 g
Fiber:  2.5 g
Sugar:  2 g
Protein: 1.6 g

As with other recipes throughout this book, added salt is fine if you can fit it into your overall sodium bank.  Since this recipe only makes 4 servings, I would limit the added salt to no more than ¼ teaspoon, but ⅛ teaspoon would be much better – taking the sodium content of the dish to 87 mgs. The ¼ teaspoon would take it to 160 mgs.

## Egg and Cheese Fritatta

This is the perfect Sunday brunch dish. It is attractive and the flavor can be changed in many different ways. The recipe is for my standard "go-to" frittata, the basic one which I can confidently serve to most people and get a favorable reaction.

### Ingredients:
6 eggs
6 servings of egg substitute
⅓ cup milk (I use fat free)
1 cup chopped onions
1 cup chopped spinach
2 Tablespoons olive oil
4 oz. low sodium goat cheese or chevre (around 60 mgs. per ounce)
Seasonings – rosemary, thyme, basil, garlic powder, black pepper, cumin – to taste

### Directions:
Pre-heat oven to 325°.

Choose an oven–compatible, medium sized skillet with two inch sides or higher. Place on stove top to start.

Add oil, onions, spinach and seasonings; cook for several minutes on medium heat.

Beat eggs, egg substitute and milk in a bowl and then add to the skillet.

Cook over medium heat for 10 minutes, covered.

Uncover and scatter the cheese over the top of the frittata.

Replace cover and move skillet into 325 ° oven and cook for another 30 minutes.

Check contents for doneness after 20 minutes – frittata should be firm when jostled, and the cheese should have melted and spread into the frittata.

Remove skillet from oven and let stand for 10 to 15 minutes.

Transfer frittata onto a serving plate and slice like a pizza into eight slices.

Serve hot.

**Nutrition Information for 1/8 of pie:**
Serving Size: 1/8th of recipe
Sodium:  199 mgs.
Calories:  176
Total Fat:  11.6 g
Saturated Fat:  2.5 g
Cholesterol:  166 mgs.
Carbohydrates:  4.1 g
Fiber:  1 g
Sugar:  2.5 g
Protein:  13.5 g

I have tried this recipe using all egg substitute – 12 servings -- virtually eliminating all of the cholesterol, and a lot of the fat in this recipe.  However, this would add another 38 mgs. of sodium per serving, taking the total up to 237 mgs.  That is not too bad, especially since I have to watch the cholesterol.  The taste is basically the same as well. Given the greater flexibility in my brunch sodium bank, the added salt that I mentioned in several of these recipes can be reasonably worked into my menu.

## Oatmeal

Oatmeal and other hot cereals are an easy and convenient item for breakfast or brunch. But I want to provide a reminder about the instant version of oatmeal and some other hot cereals. Instant packets tend to have added sodium along with the other flavorings, while the regular cooking versions of these cereals do not contain any sodium at all. What a difference! Not only does the cereal taste better if you spend a few minutes to cook it, you also get the bonus of no added salt. You can then decide how to flavor the cereal on your own, or better yet, place the toppings and additions in separate bowls and allow your guests to pick and choose. Think of all the very low or sodium free items that work well with hot cereal – honey, brown sugar, maple syrup, nuts, dried fruit, fresh fruit and berries, cinnamon, nutmeg, allspice, low sodium butter spread, milk, cream and so forth.

## Eggs with Grits

I like to combine an egg dish with grits. It is so satisfying in the morning. I make plain grits in water, and add any combination of seasonings that strike me at the moment. My egg dishes are typically made with egg substitute, so there is about 115 mgs. of sodium for each egg equivalent. Two eggs plus cheese is already about 300 mgs, so there is still some room left in the sodium bank for the meal. The cheese is certainly optional if you are sampling other goodies from the brunch menu. I will make the grits fairly spicy, and then spoon them right on top of the eggs, adding a splash of Tobasco™ sauce as well. With juice and coffee, this is a good meal in itself. For brunch, I would take a smaller portion of this dish – say half – and still have room in the sodium bank to fill a plate with several other reduced portion dishes. Another idea would be to just add the grits on top of the frittata described above, which is another combination of egg and cheese, and omit this egg dish from the menu.

## Corn Bread Muffins

Corn bread recipes are just plain salty, but corn bread is a wonderful, earthy, comfort bread for many people. Rather than consuming 250 mgs. or more of  sodium per slice or muffin, the following recipe weighs in at about 32 mgs. of sodium. My taste-tester friends uniformly say that they don't miss the sodium at all, and would not even suspect that the bread was modified in any way. That's pretty good, given how much I have modified the traditional recipe to make this one taste non-modified.

### Ingredients
¾ cup corn meal
1 ¼ cups all purpose flour
½ cup egg beaters or 1 egg beaten
1 cup fat-free high protein milk

¼ cup canola or other vegetable oil
¼ cup sugar
1 teaspoon active dry yeast
1 teaspoon Featherweight™ or other brand of no salt baking powder
Optional 1 teaspoon of vital wheat gluten

**Directions:**
Pre-heat oven to 375º.

Mix all dry ingredients together in a mixing bowl.

In small bowl, add 2 tablespoons warm water to yeast and 1 teaspoon of sugar. Let stand until small bubbles appear.

Add all wet ingredients into the dry, including the bloomed yeast.

Mix gently to incorporate all ingredients. *Do not over mix.* A few small lumps are fine.

Spoon mixture into a greased nonstick muffin pan (Pam™ or other cooking spray is fine) and fill two-thirds to three-quarters of the way to the top. This should make between 10 to 12 muffins. (I usually get a yield of 11 muffins).

Place in the middle rack of a 3 rack oven, or the bottom rack of a 2 rack oven, and bake for 18 minutes or until the muffins begin to turn a light brown.

Remove from oven and turn out onto cooling rack.

**Nutrition Information for one muffin (out of eleven) with gluten added:**
Serving Size: 1 muffin
Sodium: 32.7 mgs.
Calories: 155
Total Fat: 5.2 g
Saturated Fat: 0.3 g
Cholesterol: 0.4 mgs.

Carbohydrates: 22.2 g
Fiber: 1.1 g
Sugar: 5.3 g
Protein: 4.5 g

The optional vital wheat gluten will add a slightly chewier texture to the muffin, if that is what you prefer. Applesauce can also be used as a replacement for some or all of the sugar. Use ½ cup of applesauce for the ¼ cup of sugar. Honey is another good replacement for table sugar, as it will add a bit more moisture to the mix. Use 1 tablespoon of honey with ⅓ cup unsweetened applesauce. Corn kernels, crushed corn or cream corn (all no salt added) can also be used to add more texture to the bread. Use ½ cup for the recipe. The corn will add a small amount of calories to each muffin. But if you remove all but 1 teaspoon of sugar to bloom the yeast, and then use the applesauce and honey as replacements for the sugar, you can end up at the same number of calories, but fewer grams of sugar, and a bit more fiber and protein. These muffins are not especially sweet but they do have good corn flavor. Experimenting is, once again, the best way to nail down the recipe and taste that works for you.

# XVII.
# In the Restaurant

There have already been several references to restaurant situations elsewhere in this book. But this chapter will condense a number of tips and suggestions that can help you get through almost any restaurant experience. I enjoy a meal out with my wife and daughters, or with friends and other family members. Then there's the business meals, where I might have less flexibility to select the establishment. I don't want dining out to always be a huge challenge, and I never want to embarrass my dining party. Nor do I want to eat so blandly that I'd rather not even go out in the first place!

First and foremost, the one activity that pays off better than almost anything else you do, is to look up the menu online in advance of any trip to a restaurant and take a few minutes to browse what is available. This will help you narrow down your choices and prepare to ask some salt-related questions of the staff. If you are considering several different restaurants to decide where to go, this is a great way to determine which place will be best for you.

Most restaurants will have at least some fresh, non-brined, non-marinated protein sources available – typically fish and seafood, perhaps chicken or beef. If you call ahead and request this information, it is very possible that they will be happy to set aside a specific protein source for you. In some cases, you may need to make a reservation in order to ensure that they do this. One caveat, however – a restaurant may receive a shipment of chicken breasts, for example, which already contain a high amount of sodium per serving right out of the box. If they agree to set aside a portion for you and not marinate it, you may still end up with 350 to 500 mgs. of sodium per serving when this is the case. What luck! It's better than 900 mgs, but it's still not what you want. At least some restaurants will try to determine for you whether this is the case, while others say that they simply don't have the information and cannot do so. But you won't ever find out if you don't ask. I feel strongly there is no sodium related question that is off limits to ask of any restaurant. Most managers are more than happy to try to help you, and want you to say good things to your family and friends about the level of service and accommodation that you experienced at their establishment. The pricier the restaurant, the more that this is generally the case. But I have found really eager assistance at all price points.

Certain side dishes and accompaniments are easy to modify, while others are nearly impossible. Here is a general rule of thumb. Rice, beans, pasta, mashed potatoes, soup, breads and most appetizers are going to be difficult to modify. Most of these items are pre-made in large batches. It is not reasonable to expect that the restaurant will make a serving of rice for you without salt, for example. Salads, vegetable side dishes and main protein sources are going to be the easiest to attempt to modify. Desserts are likely to be less of a sodium problem than other non-modified portions of a restaurant meal. Here are some examples:

Most vegetables can be grilled in olive oil without any salt. Asparagus, green beans, onions, peppers, broccoli, carrots, zucchini, mushrooms, tomatoes and other vegetables can taste great when grilled this way. Even sliced potatoes can be grilled in this fashion. Many restaurants do not seem able to throw some spices on the grill since their spices are pre-mixed with salt. Bring your own! A small container of your favorite seasonings will finish the grilled vegetables tableside, so to speak. The same is true for meats, poultry, fish, and seafood. I am sprinkling at restaurants all of the time. I have never been told that I cannot bring my own spices to the restaurant, and the wait staff has seen me in action. This tip should be of help in most types of restaurants and most cuisine choices – steakhouse, Tex-Mex or Southwest, diner, even high brow continental establishments.

Most restaurants will offer you oil and vinegar on the side for your salad. I prefer that they separate the two ingredients into two small bottles or bowls and let me decide how much of each I want to add. Be careful – at some establishments, when you say "oil and vinegar on the side" what they hear is *vinaigrette on the side*, and many vinaigrettes are made with salt. I have had this problem on a number of occasions. If this happens to you after making your request, make them take the vinaigrette back and give you what you asked for. If you happen to prefer first cold pressed Spanish olive oil and ten year aged dark balsamic vinegar from a particular farm in Modena, Italy, then by all means bring your own. Again, this should not cause a problem with the restaurant. But if you can be satisfied with whatever oil and vinegar that they provide you, then perhaps just a sprinkle of the spices from home is all that is needed to make a perfect salad dressing.

Don't care for oil and vinegar?  I have ordered creamy ranch on several occasions, and used it very sparingly. Thousand Island dressing is probably about the same as the ranch dressing. Figure that these types of dressings will contain around 120 to 150 mgs. of sodium per tablespoon. It's probably a better choice than a parmesan laden dressing. Blue cheese is probably a slightly lower sodium alternative to parmesan.  Avoid Italian dressings since most have a lot of salt.  Avoid any dressing with soy sauce as one of the ingredients.  These dressings are likely to contain at least 200 to 250 mgs. of sodium per tablespoon. That knocks out most Asian-inspired dressings.

Also avoid the following items on a salad – pickles, pickled vegetables, capers, olives, anchovies, croutons, garlic bread or toast, cheese and bacon bits.  I find it's best to go over this list with the wait staff and make sure they understand what you mean by no salt added.  They often don't think about these salad additions as being highly salted.  It may be on your mind all of the time, but that doesn't mean it's on everyone else's mind.  The more you communicate, the better.  It may take an extra minute to order your meal, but you will be better off for it.  My family used to cringe when I started reviewing my medical history with the waiter, but I'm better now, and I try to accomplish this in a more workmanlike fashion.

Bring one of your homemade pan-Asian stir fry sauces with you to any Asian restaurant and you should be able to fare well with a steamed main course of protein, vegetables and rice. Your homemade peanut sauce can even go well with fresh spring rolls wrapped in rice paper. While large sections of the Asian menu may be off limits, at least you will be able to join family and friends and enjoy yourself rather than feel deprived. If you can't stand steamed food no matter how you might try to dress it up, then you can try ordering a stir fry with oil only, but good luck with this! You may get what you want some of the time. You have to be very explicit, and very patient, since you might find yourself waiting for a second try at your meal while everyone else starts eating.

I have had very little luck at the typical pizza parlor. There isn't any way to get the restaurant to modify the pizza crust, tomato sauce, cheese, meat, and other toppings that would result in a meaningful sodium reduction. So save up your sodium and have one slice of pizza with a big salad (most of these establishments offer salad choices these days), and savor every bite of the fully loaded pizza! But avoid additions like pepperoni and sausage since these just add lots more salt.  Don't forget to manage the salad as discussed earlier in this chapter, focusing on both the ingredients and the salad dressing.

Staying in the Italian realm, even if the pasta is boiled in salted water, a good, moderate sodium dish might include fresh pasta, grilled tomatoes and other vegetables, a grilled protein that has been set aside for you, fresh garlic and good olive oil.  Most places can work out something along these lines to your satisfaction.  Bring your own Italian seasonings just in case they cannot season your dish given a pre-mix with salt.  While bread will be salty, a small piece dipped in olive oil with your home seasonings is not a bad addition.  If they happen to make a Tuscan-style bread, it will likely not contain any sodium.  It's worth asking the question.

Almost any type of restaurant can grill fresh fish and not use any salt.  Fish may be the easiest choice in most eating establishments.  But it should be fresh, not frozen.  Frozen fish is more likely to have added salt as part of the preservation process.  Certain fruits of the sea are almost always salted when frozen, such as crab legs, so ask the questions and be persistent without becoming too enormous of a pest.

A dry baked potato is a great choice for a side dish, provided the restaurant can assure you that they did not salt the outer skin as part of the baking process. Good low sodium additions to a baked potato are sour cream, chives or scallions, and even a small amount of butter. But stay away from barbeque sauces, cheese and bacon bits. A baked potato is filling and when combined with a good steak and a properly crafted salad, your restaurant experience might approximate the same milligrams of sodium as the home cooked version of the meal. But somehow, it always tastes better when dining out.

# XVIII.
# Final Thoughts

This journey is never over.  There are new products, new recipes and new ideas entering the scene practically all of the time.  The pace of change may be a bit slow for some, but change is here.  The future seems destined to be a lower sodium world, or at least one where the options are far more significant.  But many people cannot wait for that future to materialize.  The pace with which the major food manufacturers and restaurants are removing sodium from foods is unacceptably slow if you have already been told by the doctor that the time is now to cut back to 1,500 or 2,000 mgs. per day.  There are plans by several major food manufacturers for a 10% reduction in sodium content over the next year or so, with further plans for another 10% cut down the road.

Companies are moving slowly for a good reason – they want to subtly alter the sodium content so that the average salt encrusted taste bud does not go into violent revolt against the progress. I have no problem with that. However, at the same time that they are tippy-toeing the sodium reduction with existing food items, they could be far more aggressively introducing new items that are dramatically lower in sodium. But marketing doesn't seem to work that way – the greatest effort is made for the greatest market, and the market for very low sodium foods, while growing, is still quite small compared to the mainstream demand. Perhaps an increased focus on lower government guidelines for sodium will help to ramp up the pace of change.

It is clearly up to the individual to take this challenge by the horns. This book is an assemblage of nearly five years worth of information gathering, learning experiences, and experimentation. It shows that not only can a low sodium lifestyle be achieved, it doesn't have to taste like cardboard and be boring and depressing. Far from it! While we each have the power to make this happen in our own lives, to some extent it is like swimming upstream -- you can move through the water, but the current is definitely against you. It is easier not to try, especially if you do not have a history of cooking and baking. In this microwave ready and fast food restaurant world, that would be an apt description for a large number of people. It would be easier to reach for products containing the words "lower sodium," which are showing up more and more, and just settle for that, even though it may make the 1,500 to 2,000 mgs. per day goal nearly impossible to reach. But it would still be better than doing nothing.

People ask me all of the time, "How do you find the time to do this?" I made sure to mention many time saving techniques in this book as an answer to those questions. People ask, "Where did you learn?" That answer is more complex – I am not afraid to try and fail, and I will listen, watch and learn from any source I can find, be it television, the internet, talking to other people, or just striking out on foot and checking what's new in the stores.

I wished for a book that made an effort to put it all together, but I could never seem to find one. There are low sodium cook books with lots of recipes, but I wanted much more than that. I needed to figure out how this was going to work on a day by day basis, and I needed a bigger picture approach to the whole subject rather than just learning individual recipes. So I wrote Little Crystals to be about all of these things. I tried to talk about the how, the why, and the what on the subject of low sodium living. There are thousands of low sodium recipes available beyond the scope of this book, but once a reader comes to appreciate the ability to develop a personal framework for managing a low sodium lifestyle, it will be easier to deal with all of those recipes out there in bookstores and on the internet. Concepts like the sodium bank and seasoning the pot, workhorse ingredients, flavor combinations, the ability to make a good salt free bread product, and developing the knack for substitutions will all help to give a person more of a sense of control over the changes that have to be made. Creating a personal list of "do not disturb" items, which offer your particular taste buds the most bang for the sodium buck, and learning how to incorporate those items into a low sodium daily meal plan, will give you a sense of satisfaction that makes sticking to this lifestyle just that much easier.

One more helpful idea needs to be mentioned before the close of this book. When you make your own dishes through variation and experimentation, it's pretty easy to lose sight of the nutritional details of all of the ingredients. The easiest way to keep track, assuming you possess basic computer skills, is to create a spreadsheet which allows you to input all nutritional information for the whole of the recipe. All sodium from all ingredients, plus all calories, protein, sugar and so forth, can be accurately tracked. Use the nutrition labels on your products to gather the needed information.  For produce and other items, you can also check the government website *nutritiondata.com* for a detailed nutritional breakdown of virtually every food product under the sun.  There are many other websites that provide this information, including some from the manufacturers themselves. There are other resources as well, such as smart phone applications that contain much of this information.

As shown on the chart below, total the columns and then add one more line where you divide the totals by the number of servings that you plan to make from the total recipe.  Sometimes, I add a few such lines, to see the impact of dividing by 8, let's say, versus dividing by 9 or 11 servings.  The point of this exercise is to figure out just how many healthy portions you are making when you make a large pot of something or a large casserole dish.  If the protein and sodium look right only when you divide by a higher number, say 10 portions, yet you feel the meal looks a bit skimpy when divided by that high of a number, this may be a signal to add more vegetables to the pot, for example.  I will make adds and take-aways until I am satisfied with the look of the dish, the size of it, and of course, the nutritional profile and sodium levels.  Accordingly, some of my recipes make more than the standard 8 servings, which is fine with me.  A few make less than 8, and that is also fine once I understand the details to arrive at that decision. Here is an example of how this works with the brunch item, mejadrah, a recipe that could be served in differing portion sizes depending upon what else is on the menu for a given occasion:

| | SERVING SIZE | SODIUM | CAL | TOT FAT | SAT FAT | CHOL | CARB | FIBER | SUGAR | PROTEIN |
|---|---|---|---|---|---|---|---|---|---|---|
| **Mejadrah** | | | | | | | | | | |
| Onions - sliced | 4 cups | 26 | 256 | 0 | 0 | 0 | 60 | 12 | 28 | 8 |
| Green lentils | 1 cup dry | 12 | 678 | 1.9 | 0 | 0 | 115.4 | 57.6 | 3.9 | 48 |
| Jasmine or basmati rice | 1 cup dry | 0.0 | 640.0 | 0.0 | 0.0 | 0.0 | 144.0 | 4.0 | 0.0 | 12.0 |
| Olive Oil | 1/4 cup | 0.0 | 504.0 | 56.0 | 4.0 | 0.0 | 0.0 | 0.0 | 0.0 | 0.0 |
| Salt | 1/2 teaspoon | 1180.0 | 0.0 | 0.0 | 0.0 | 0.0 | 0.0 | 0.0 | 0.0 | 0.0 |
| | Total: | 1218.0 | 2078.0 | 57.9 | 4.0 | 0.0 | 319.4 | 73.6 | 31.9 | 68.0 |
| | 1/8 serving | 152.3 | 259.8 | 7.2 | 0.5 | 0.0 | 39.9 | 9.2 | 4.0 | 8.5 |
| | 1/10 serving | 121.8 | 207.8 | 5.8 | 0.4 | 0.0 | 31.9 | 7.4 | 3.2 | 6.8 |
| | 1/14 serving | 87.0 | 148.4 | 4.1 | 0.3 | 0.0 | 22.8 | 5.3 | 2.3 | 4.9 |

Sticking to your sodium target is what this book is all about. When people are told to cut salt, it usually means for the rest of your life. That may sound like a prison sentence to some, but attitudinal change, coupled with knowledge, can greatly soften the blow. If a reader sticks to the low sodium approach for the long haul, and makes sure to get adequate daily protein, watches fat and sugar in their meals, drinks lots of water and makes sure to get plenty of fiber from fruits, vegetables and beans, then over time, most people should feel better, reduce or eliminate the symptoms of diagnosed sodium sensitivity, reduce high blood pressure, help support their kidneys, minimize calcium depletion in bones, look better, perhaps lose weight and reduce any edema that might be affecting their limbs. What better payoff is there to giving something up (excess salt) than getting something in return – a better, healthier you!